# OUTLOOK FOR NATURAL GAS—
# A QUALITY FUEL

*The Proceedings of the Institute of Petroleum Summer Meeting,
"Natural Gas — Outlook for a Quality Fuel", held at the
Palace Court Hotel, Bournemouth, 6—9 June 1972*

# Outlook for Natural Gas— a Quality Fuel

Edited by
**PETER  HEPPLE**

A HALSTED PRESS BOOK

JOHN WILEY & SONS
New York — Toronto

PUBLISHED IN THE U.S.A. AND CANADA BY
HALSTED PRESS
A DIVISION OF JOHN WILEY & SONS, INC., NEW YORK

The symbol I.P. on this book implies that the text has been officially
accepted as authoritative by the Institute of Petroleum, Great Britain

**Library of Congress Cataloging in Publication Data**

Main entry under title: "Outlook for Natural Gas".

A Halsted Press book.
Proceedings of the Institute of Petroleum Summer Meeting...
held at the Palace Court Hotel, Bournemouth, 6–9 June 1972.
1. Gas, Natural – Congresses. 2. Gas as Fuel – Congresses.
I. Hepple, Peter, ed. II. Institute of Petroleum, London.
TN880.O87    1973    333.8'2    73–661
ISBN 0–470–37303–2

WITH 58 ILLUSTRATIONS AND 38 TABLES

ⓒ APPLIED SCIENCE PUBLISHERS LTD 1973

Printed in Great Britain by Adlard & Son Ltd
Bartholomew Press, Dorking, Surrey

# CONTENTS

# LIST OF CONTRIBUTORS

B. BONNETT
Exploration and Production Manager, Santos Ltd, Adelaide, Australia

D. G. M. BOYD
Head, Market Research and Evaluation Division, Shell International Gas
Ltd, Shell Centre, London

H. J. BUCKLEY
Engineering Manager, Imperial Continental Gas Association,
2 Devonshire Square, Bishopsgate, London

S. E. CHURCHFIELD
Group Manager, Exploration and Production, Burmah Oil Trading Ltd,
London

P. F. CORBETT
Marketing Services Manager, Gas Division, International Sales Department,
British Petroleum Co. Ltd, London

B. DAVEY
Manager, Process Engineering, Air Products Ltd, New Malden, Surrey

R. C. FFOOKS
Naval Architect, Conch Methane Services Ltd, London

A. F. FOX, M.B.E.
Group Chief Exploration Manager, Tricentral Ltd, London

A. I. D. FRITH
Marketing Manager, Domestic and Commercial Gas, The Gas Council,
London

M. D. J. GELLARD
Head of Energy, Economies and Economics, Gross Division, Shell
International Petroleum Co. Ltd, Shell Centre, London

I. A. I. GRIFFITHS
Director, Operations, Shell International Gas Ltd, Shell Centre, London

J. HAMMOND
Gas Development Manager, Corporate Developments, Burmah Oil
Trading Ltd, London

W. HENDRY
Sales Manager, Air Products Ltd, New Malden, Surrey

C. HIJSZELER
Head of Gastechnological Research and Advice Department,
N.V. Nederlandse Gasunie, Groningen, The Netherlands

D. H. NAPIER
Senior Lecturer, Department of Chemical Engineering and Chemical
Technology, Imperial College, London

M. W. H. PEEBLES
   Manager of Planning and Finance Division, Shell International Gas Ltd,
   Shell Centre, London
N. W. ROBERTS
   Business Development Manager, CJB (Projects) Ltd, London
J. M. SOESAN
   Commercial and Legal Director, Conch Methane Services Ltd, London
G. G. SPALDING
   Staff Engineer, Amoco (U.K.) Exploration Company, London
M. J. STEWART
   Process Engineering Department, CJB (Projects) Ltd, London
R. M. THOROGOOD
   Manager, Process Technology, Air Products Ltd, New Malden, Surrey
C. TUGENDHAT, M.P.

# World Review of Natural Gas Sources

## By CHRISTOPHER TUGENDHAT, M.P.

It is a pleasure to be invited to give the keynote address to this Summer Meeting because I have been connected with the oil industry throughout my working life. Indeed, I notice several people here whom I interviewed when I was on the staff of the *Financial Times*. I have now moved into another sphere, but I am still very interested in the oil industry as a Member of Parliament and as a director, in addition to journalism.

Since I first began writing about oil, in the early 1960s, the growth rate has been explosive and huge changes have come about. Exploration is now centred on the offshore areas; vast new oilfields are coming into production, notably in Alaska, the North Sea and Australia; and we have seen the development of OPEC.

But there has been nothing more impressive than the growth of natural gas, which bids fair to become the third important phase of the oil industry's history. The first was the discovery and utilization of oil itself; then the emphasis moved towards petrochemicals, in the development of which my own father played an important role; now we are witnessing the rise of natural gas.

As has been so often the case in the oil industry, the U.S.A. led the way, for natural gas was used widely in the States before the Second World War. By 1945 it had a 15 per cent share in the total U.S. energy market and by the 1960s this had grown to 30 per cent.

Europe lagged far behind. Soon after the war important discoveries were made in Italy and France but these were of largely local interest. It was not until the early 1960s, with the really big finds in Holland — which pointed the way to the North Sea discoveries — and Algeria, that natural gas began to make a significant impact on Europe as a whole.

The first natural gas was imported into the U.K. from Algeria in 1964, the same year that natural gas sales began in Holland. In 1971, the fuel had taken 30 per cent of the total Dutch energy market and it will probably have a 45 per cent share by 1975. Within six years it is likely that the demand for natural gas will have quintupled.

Like the oil industry, the gas industry is intensely international. In Europe there are large-scale exports of gas from Holland to neighbouring

1

countries and imports from North Africa will go to Spain and Italy as well as existing markets in Britain and France. There are also developments further afield, from Alaska to Japan, for instance, from Algeria to the U.S.A., and from Brunei to Japan.

But what we have had up to now is a mere *hors d'oeuvre* — the real growth is still to come. One table (Table I) is sufficient to show the scale of natural gas reserves. Between the end of 1972 and the 1980s, international movements of LNG should increase more than 20 times. At this year-end the total will be about 2 billion cu ft/day (using billion as 1000 million): by the mid 1980s it is likely to be 40 billion cu ft/day.

The object of a keynote address is to establish the framework for a conference. I have already given you some idea of the scale of the industry we are talking about. For the remainder of this address I shall try to answer the following four questions:

1. Why will the natural gas industry grow?
2. Where will natural gas be needed?
3. Where will it come from?
4. What will happen to the price?

The first question is easy to answer, for the non-Communist world is becoming increasingly worried about energy supplies. Nuclear power has not lived up to expectations — even in Japan it is estimated that it will represent no more than 13–14 per cent of the total energy market in 1985. The coal industry is declining rapidly — faster, in my opinion, than has been supposed. Oil can fill part of the gap and is growing fast, but suffers from political unreliability. Therefore, natural gas can play a vital role.

There is another reason, too, and that is the growing concern over pollution and the environment. It is notable that in my book *Oil: The Biggest Business*, written only in 1968, there is no mention of pollution and I was never criticized for this omission, whereas at this very moment in Stockholm there is taking place a large international conference on the human environment. Natural gas is the cleanest of the fossil fuels and this fact will undoubtedly become more important.

The short answer to the second question is "everywhere". The biggest buyer will be the U.S.A., for by 1980 the demand is expected to exceed the domestic supply by 25 billion cu ft/day. Tanker shipments of LNG should provide at least 5 billion cu ft/day of this.

Japan is also a very big market. The Energy Council there recently forecast the LNG imports to that country will reach 3 billion cu ft/day by 1985.

Despite the production from Holland and the North Sea, Western Europe's needs will grow. Within the next four years a billion cu ft/day will be imported and this will be doubled by the end of the decade. In addition to the existing suppliers, one can expect Russian gas to come in by pipeline.

## TABLE I

## Gas Reserves by Country

Figures as from 1 January 1972 (as published in *Oil and Gas Journal*, 17 April 1972)

| ASIA – PACIFIC | Billion cu ft |
| --- | --- |
| Afghanistan | 4,900 |
| Australia | 24,800 |
| Brunei – Malaysia | 7,500 |
| Burma | 100 |
| India | 1,500 |
| Indonesia | 4,500 |
| Japan | 400 |
| New Zealand | 6,000 |
| Pakistan | 19,500 |
| Taiwan | 600 |
| Total | 69,800 |

| MIDDLE EAST | |
| --- | --- |
| Abu Dhabi | 10,000 |
| Bahrain | 5,000 |
| Dubai | 1,000 |
| Iran | 200,000 |
| Iraq | 22,000 |
| Israel | 60 |
| Kuwait | 35,000 |
| Neutral Zone | 8,000 |
| Oman | 2,000 |
| Qatar | 8,000 |
| Saudi Arabia | 52,000 |
| Syria | 700 |
| Turkey | 170 |
| Total | 343,930 |

| EUROPE | |
| --- | --- |
| Austria | 550 |
| Denmark | 500 |
| France | 6,900 |
| Germany, W. | 14,000 |
| Italy, Sicily | 6,000 |
| Netherlands | 83,000 |
| Norway | 10,000 |
| Spain | 500 |
| United Kingdom | 40,000 |
| Yugoslavia | 1,800 |
| Total | 163,250 |

| AFRICA | Billion cu ft |
| --- | --- |
| Algeria | 106,500 |
| Angola | 1,500 |
| Egypt | 7,500 |
| Gabon | 6,500 |
| Libya | 29,500 |
| Morocco | 18 |
| Nigeria | 40,000 |
| Tunisia | 1,500 |
| Total | 193,018 |

| WESTERN HEMISPHERE | |
| --- | --- |
| Argentina | 7,600 |
| Bolivia | 5,000 |
| Brazil | 5,000 |
| Chile | 2,200 |
| Colombia | 2,500 |
| Ecuador | 6,000 |
| Mexico | 11,500 |
| Peru | 2,500 |
| Trinidad, Tobago | 5,000 |
| Venezuela | 25,400 |
| U.S.A. | 269,596 |
| Canada | 54,376 |
| Total | 396,672 |

| SOVIET BLOC | |
| --- | --- |
| Russia | 546,000 |
| China | 4,000 |
| Hungary | 3,000 |
| Poland | 5,000 |
| Total | 558,000 |

| TOTAL WORLD | 1,724,670 |
| --- | --- |

The U.S.S.R. will be a major force in the world natural gas market. As the table shows, it contains around 30 per cent of total gas reserves, a larger proportion than any other country. Moreover, some of its largest fields are still unexploited. Production conditions are so difficult that outside help may be needed, but nobody doubts that one way or another Russia will be a major exporter. To some extent it will also be an importer as well. It is already bringing in supplies from Iran and Afghanistan. But this is simply because the fields in those countries are more conveniently situated to supply the southern part of the Soviet Union than Russia's own fields. The imports will enable it to export much more than would otherwise be the case to the West. There is a strong possibility that some at least of the exports will go to the U.S.A., perhaps in a barter deal involving the exchange of U.S. grain for Russian gas.

As the position of Russia shows, gas reserves do not follow the same pattern as those for oil. They frequently occur in the same place, but not always. As the Groningen field and the Southern North Sea have demonstrated, natural gas may be found alone. Although the Middle East accounts for 55 per cent of the world oil reserves, the figure for gas is far less, being only around 20 per cent.

The final question, about what will happen to the price, is also capable of being given a short answer. It will go up, the real point being by how much? The rise will occur worldwide, though there will be national variations depending on the extent of government intervention and market conditions. But in an international market in which the U.S.A., Western Europe and Japan are competing for supplies, some form of international price structure is bound to emerge.

The writing, in fact, is already on the wall. The landed price in Japan of gas from Shell's Brunei field will be 48·6 U.S. cents per 1000 cu ft but in a new contract signed a few months ago the figure is believed to be 80 cents. In the U.S.A., prices at the wellhead range from 22·5 to 26 cents per 1000 cu ft but the contract delivered price at New York City is 42 cents. Import estimates range from 60 to 90 cents.

This brings me inexorably to the North Sea, where I should like to finish my talk. The Gas Council has enough reserves to support its expansion plans up to the late 1970s, but thereafter it will need more. This will be so even if it wants only to maintain the level of consumption then reached and even more so if it wishes to continue to expand.

The first place to look towards is the North Sea itself, from which supplies are at present drawn from the southern part. But more natural gas has been discovered in Norwegian waters and in the Frigg field which straddles the boundary. These finds are part of the international market. The present price paid for North Sea natural gas is just over 31·2 U.S. cents per 1000 cu ft. Although comparisons are difficult, the international trend is clear. Prices for new North Sea gas will have to rise considerably above present levels.

Nobody likes paying more, but in my opinion it will be worth it. The U.S.A. is now paying the penalty of shortage for holding down its internal prices. If we underprice North Sea gas, this valuable commodity will not be available in sufficient quantities to meet our needs.

# Historical Review of Some Equipment and Techniques Available to the Natural Gas Industry

By B. BONNETT, S. E. CHURCHFIELD and J. HAMMOND

*(Burmah Oil Trading Ltd)*

## SUMMARY

This paper reviews the history of six of the technical facets of the natural gas industry that have had a major influence on its development, these being reflection seismic survey, deep drilling, offshore drilling, gas processing, pipelining, and LNG.

Some of these techniques are common to, or closely associated with, those used in the oil industry, a notable exception being the recent developments in the field of LNG. It is concluded from the past performance that the industry will meet future technical challenges given adequate economic incentive.

Seepages of natural gas have been known since earliest historical times. The Chinese are reported to have piped gas through hollow bamboos in the Shu Han dynasty, while in 1821 hollow logs were used to transport natural·gas from shallow wells to light the streets of Fredonia in New York State.

Drake's well, drilled at Titusville, Pennsylvania in 1859, usually considered to be the start of the oil industry, can also be regarded as the start of the natural gas industry, since within a few years gas was being produced and piped for sale.

Up until recent times, oil has been the primary objective of exploration and all the major gas provinces throughout the world were discovered in the course of the search for oil.

The techniques of exploration for oil and for gas are identical; much natural gas is produced in association with oil, and there are many similarities in equipment and techniques between oil and gas fields.

Historically, gas has been less valuable and is less easily transportable compared with oil, and whether associated or non-associated with oil its exploitation and sale has depended upon the availability of a market and within a technically and economically feasible distance. Pipelining, therefore, has been a critical factor in the development of the natural gas industry, and here again there is much similarity between equipment and techniques used to lay and operate oil and gas pipelines.

Recovery of natural gasoline and treatment of associated gas for pipeline use is an integral part of the operation of many oilfields, and although natural gas processing has developed independently, there are many associations with the oil production and refining industries.

In contrast to exploration, production, pipelining, and processing, the equipment and techniques used in the most recent major development in the natural gas industry – the manufacture and transportation of LNG – are largely independent of the oil industry. The recent developments in the field of manufactured gas (SNG), while potentially of great future significance in the overall energy pattern, are considered to be outside the scope of this paper.

In subsequent sections of this paper the authors have attempted to outline the history of development of some equipment and techniques which they consider are relevant to the present and future state of the natural gas industry.

The criteria leading to the selection of the subjects are:

That the reflection seismic survey as the main source of pre-drilling subsurface geological information will remain a major tool in the exploration for petroleum (including natural gas);

That the continued development of techniques for drilling to greater depths and in offshore areas has been and will be a major factor in discovering and exploiting new gas accumulations;

That processing, and particularly pipelining techniques, have been critical in establishing natural gas as a major energy source in many areas of the world;

That this position will become worldwide by the transportation of liquefied natural gas.

The authors appreciate that their choice of subjects is controversial, and due to limitations of space they are not treated in the detail their importance merits, and that many which others may correctly feel are more significant have been left out.

## EXPLORATION GEOPHYSICS

The major part of reserves of petroleum to be discovered in the future will be on the continental shelves, in isolated or inaccessible onshore basins, in horizons deeper than those presently explored, and in areas where geological complexities have not been unravelled by existing exploration techniques. Geophysics and, in particular, reflection seismic surveys, will be the main pre-drilling exploration tool used to find these reserves.

Colonel Drake located his well in a valley close to an oil seepage and in the immediately following years creekology and surface manifestation of petroleum were the primary criteria used to locate exploration and extension wells.

The relation between oil seepages and anticlines was noted by Sir William Logan of the Canadian Geological Survey as early as 1848, and T. Sterry Hunt, also of the Canadian Geological Survey, formulated the anticlinal theory of accumulation of petroleum in 1861, but the theory was not accepted at the time as the accumulations in what was then the centre of activity in Pennsylvania are controlled by stratigraphy rather than structure. I. C. White revived the theory in the 1880s and successfully used it to find gas in West Virginia, and in the next 50 years surface geological mapping to locate anticlinal features was the main scientific exploration technique.

The concept of using artificial seismic waves for subsurface exploration was presented by Robert Mallet in a paper to the Royal Irish Academy in 1846. Experimental and theoretical studies were carried out in the following 70 years, and following the boost given by developments related to location of enemy gun positions during the First World War, a patent covering the use of refraction profiling for locating subsurface formations was applied for in 1919 by the German scientist Mintrop.

German seismic refraction crews started work in Mexico in 1923 and in the Gulf Coast in 1924 where, using the fan or arc shooting techniques, they were highly successful in locating salt domes. Refraction shooting was also used successfully during the 1930s in the Middle East, where geological conditions were favourable, but elsewhere results were frequently unsatisfactory and the method was replaced by the faster and cheaper siesmic reflection technique. Currently, use of refraction shooting is largely confined to problems not susceptible to reflection shooting.

The concept of reflection shooting was developed from sonic water depth sounding techniques and the first patent covering its use for mineral exploration was applied for in the U.S.A. in 1915. The method was developed in the U.S.A. during the 1920s and its value was fully established by 1930 following several significant discoveries resulting from its use in Oklahoma; by 1937 almost 250 seismic reflection crews were in operation.

The 1930s saw a steady improvement in instrumentation and techniques, the introduction of the automatic gain control and the development of dynamic geophones being major advances. In the 1940s there were further improvements in the design of recording equipment and considerable experimentation with geophone patterns and shooting techniques. The wartime development of underwater acoustic equipment and of position locating devices led to a rapid expansion of marine seismic surveys.

There were many innovations during the 1950s, including magnetic tape recording, visual display sections, and non-dynamite energy sources, but these were all overshadowed by the introduction of the Common Depth Point (CDP) or stacking technique. In this method, successively overlapping geophone spreads are shot so that in effect a number of records are obtained from the same reflecting point but from different shot hole and geophone positions. By combining these records, the genuine reflections are enhanced and spurious events eliminated.

The sensational improvement in the quality of evidence from reflection seismic surveys resulting from the use of stacking has permitted more definitive geological interpretation of subsurface structural and technological conditions, and as such makes this technique probably the biggest single technical advance in the history of exploration for oil and gas.

In the 1960s the introduction of digital recording and increased use of computer processing techniques resulted in further improvement in data quality, leading to more accurate interpretation of subsurface geology, and this trend will continue.

## DRILLING

The primary objective of the majority of wells throughout the world has been to find or develop oil, and consequently the development of the drilling industry has been basically a reflection of the requirements of the oil industry. Two of these developments, however, have been especially significant in relation to the gas industry — the trend towards deeper drilling and the trend towards drilling in offshore areas.

## DEEP DRILLING

The increase in temperature and pressure with depths below about 15,000 ft results in breakdown of the higher hydrocarbons in petroleum and a marked increase in the proportion of gas and gas-condensate pools. The development of techniques to drill to greater depths is therefore of particular interest to the natural gas industry, since it is specifically orientated to discovering and exploiting gas and gas-condensate reservoirs.

Up to the turn of the century, virtually all drilling was by the cable tool method and this persisted in some areas into the 1920s, and still remained a completion technique up to the time of the Second World War.

The method was best suited to hard rock areas and a record depth of 7759 ft was reached in Pennsylvania in 1925.

Long before this, however, the rotary method, which by the circulation of drilling mud to remove the cuttings and hold up the walls of the hole overcame the problems of soft rock drilling, had become established as the main drilling method. The discovery well at Spindletop in 1901 is generally regarded as the date when the method became established.

From the earliest days there has been continuous evolution in the design of all equipment used in rotary drilling. The introduction of seamless pipe in the

1920s, the use of higher grade steels and the development of thread types facilitated the setting of casing at increasingly greater depths.

The development of drilling mud technology, which initially, in the 1930s, was concerned mainly with viscosity control but subsequently led to the realization of the importance of the effect of filtrate loss on formation stability, also played a major part by improving hole conditions.

The progressive improvement of the roller or cone type rock bit, which was first used in 1908, has given faster penetration rates and more time on bottom, while the introduction of jet drilling in the early 1950s and the better understanding of the role of hydraulics also improved penetration rates and hole conditions, particularly in soft rock areas, and thus facilitated drilling longer ranges of hole without running casing.

As holes became deeper, the conditions encountered necessitated development of high-pressure well control equipment and of cements and muds, of well logging equipment, and of downhole tools capable of withstanding higher temperatures.

The success of the drilling industry in combating the problems of deeper drilling is exemplified by the Lone Star Producing 1 E.R. Baden well, which has reached a final depth of 30,050 ft and is the first well to drill below 30,000 ft. While the problems and costs multiply as the depth increases, from a technological viewpoint there seems no insurmountable reason why significantly greater depths should not be reached if there is sufficient economic incentive.

## OFFSHORE DRILLING

The continental shelves of the world provide the largest remaining potential for the discovery of oil and natural gas. The development of techniques to drill wells in offshore areas is therefore critical to the present and future development of the natural gas industry.

Designs for offshore drilling equipment were patented as early as 1869, but the first offshore operations were not carried out until the turn of the century, when in California and in the Caspian Sea wells were drilled from jetties or piers built out from the shore.

The first recorded use of an independent drilling platform was in 1911 in Caddo Lake, Texas, and subsequently timber and, later, concrete platforms were used to develop the Lake Maracaibo fields in Venezuela.

In 1933, the first controlled directional well was drilled to develop the offshore extension of Huntingdon Beach Field in California and in the same year the first submersible drilling barge started work in the bayou country of the Louisiana Gulf Coast. This barge, carrying the rigged-up drilling equipment was towed to the location and set on bottom to provide a static drilling base. In 1934 the use of barges to carry boilers, pumps, and ancillary equipment permitted a marked reduction in the size of the fixed platforms used in Lake Maracaibo.

The first successful well in offshore Louisiana was drilled in 1937 on the Creole Field from a fixed platform installed in 14 ft of water, while in Lake Maracaibo by 1940 wells were being drilled from platforms installed in up to 60 ft of water.

While the fixed platform was, and remains, a satisfactory technique for offshore development drilling, it is expensive and inefficient for exploration drilling, and in consequence a number of alternatives have been developed.

The first approach, introduced in the Gulf Coast in 1947, was a development from the barges used in Lake Maracaibo. A small fixed platform carried the derrick and drawworks, and all other equipment and consumables were carried on a drilling tender moored alongside. This was a major breakthrough and by 1954 there were 22 such rigs in operation.

Development based on the submersible swamp barges used in Louisiana provided an alternative and more mobile solution to the problem, but although the prototype, the Hayward-Barnsdall Breton Rig 20, was successfully introduced in 1948, and could operate in water depths of 20 ft, it was not until 1954 with the introduction of the Odeco "Mr Charlie" with a 40-ft water depth rating, that the concept of the offshore submersible rig was accepted by the industry. By 1957, the Kerr McGee Rig 46, incorporating vertical stabilizing colums, could drill in 70 ft of water, while the same company's Rig 54, built in 1962, can work in water depths up to 175 ft.

The limited capacity of the early submersible rigs led to the development, also in the Louisiana Gulf Coast, of the bottom-supported jack-up rig in which the drilling platform was floated to location and then raised on legs. The first rig which was introduced in 1954, could operate in water depths up to about 50 ft, but the capability of this type of rig was increased rapidly and the latest models can operate in water depths of up to 300 ft, which appears to be about the maximum water depth for this type of rig.

The first experiments in drilling from a floating vessel were performed in 1953 by the CUSS group (a combination of Continental, Union, Shell and Superior) in the Californian offshore area in water depths ranging from 300-400 ft, using a converted naval vessel, the *Submarex*. Following this success, a second vessel, the *CUSS I*, was converted and put into use in 1956. The problems of mooring and of drilling and completing wells in water depths of up to 350 ft were successfully overcome and the feasibility of operating from a floating vessel was clearly demonstrated.

Subsequently, development has followed two trends, each with advantages; the drillship, nowadays with a specially-designed hull and frequently self-propelled; and the semi-submersible rig, deriving buoyancy from columns and pontoons designed to offer minimum resistance to wave action.

Both drillships and semi-submersible rigs are capable of drilling to 20,000 ft and can be anchored in water depths up to about 600 ft. Dynamic positioning techniques which maintain the rig on location with computer-controlled thrust propellers have been successfully used in deeper waters.

## PIPELINES

The development of pipelines has been a critical factor in the expansion of the natural gas industry and the future of much of the industry will remain dependent on the ability to build and operate pipelines economically under increasingly adverse conditions, offshore and in such regions as the Arctic.

Much of the history of the development of pipelining techniques is common to the oil and gas industries but, in general, up to the end of World War II, whereas oil lines were laid in many countries, gas pipelines were largely confined to North America.

Both cast and wrought-iron pipe were in use for gas transmission by the 1870s, with steel pipe following by 1887. During this period pipe diameters generally ranged from 2 to 8 inches with operating pressures of 60 psi or less. The first high-pressure line operating at a pressure of about 525 psi was laid in 1891 over a distance of 125 miles from Greentown, Indiana to Chicago. 1891 was also the year in which the Dressler coupling was introduced; this coupling gave a joint much superior to those previously possible, and was to remain in general use on almost all natural gas lines until about 1940. The turn of the century saw new and improved valves and regulators coming into use and in 1910 the first long-distance (135-mile) 16-inch diameter line was constructed. This period also saw the introduction of the orifice meter which, using the Wymouth formula, enabled more accurate measurement of gas flows.

The development of large diameter seamless pipe and high yield strength steels, by permitting an increase in operating pressures up to some 1000 psi, were major factors in extending the economic length of gas pipelines which, until this time, had been limited to about 200 miles. Early lines were built using screw couplings and although oxyacetylene welding had been used as early as 1911, it was 1922 before it was accepted. The use of welding spread rapidly and the late 1920s saw the introduction of electric arc welding, which gradually replaced gas welding.

In the early days pipe had been hauled by oxen or horses and laid by manual labour. The progressive introduction of mechanical equipment for transporting and handling pipe and for ditching and back-filling was a further factor in extending the lengths of gas line. Steam-driven ditching machines had appeared as early as 1912, but mechanization made little impact until the 1920s, when track-laying vehicles started to be used. The ubiquitous bulldozer appeared in 1932 and present-day type ditchers, backhoes and drag lines were in wide use by the late 1930s.

Modern corrosion protection also largely stems from this period, a milestone being the meetings between the U.S. Bureau of Standards, ASME and API held during 1926-28. Prior to this period, efforts at protection had been confined to manual covering with cement, asphalt, priming paints and various wrapping materials. By 1930 coal tar enamel had established a predominant position as a protective material with powered cleaning and wrapping beginning to emerge, although it was the late 1930s before the present-day type of powered cleaning and priming, coating and wrapping machines evolved.

Cathodic protection, which appears to have first been employed in New Orleans in about 1928, with rectifiers first being introduced in 1930, is now standard practice on modern pipelines.

The gas-engined reciprocating compressor, introduced at the turn of the century, was to remain virtually unchallenged as the means of gas compression until the 1940s. There were, however, many improvements. The replacement of belt drive by direct drive on common frames; rise of the two-stroke engine to dominance during the 1920s, the introduction of angle designs, supercharging, etc., all resulted in greater efficiency. The reciprocating compressor was not to be challenged in its role as a gas compressor until after the Second World War. The development of centrifugal compressors, initially for single-stage low compression-ratio high-volume duty, combined with electric drives where power cost could be justified, required the development of the gas turbine before a significant impact could be made. The first gas turbine drives were introduced in about 1949, although a further decade was required to establish the new combination of centrifugal/gas turbine drive compressor.

During the period 1920-40 both oil and gas line development had much in common, with the notable exception of the much earlier use of large diameter pipe for gas transmission. Until the construction during the Second World War of the "Big Inch" and "Little Big Inch" oil lines, crude and product lines were normally limited to 12-inch diameter. The experience gained during the 1930s in the use of large diameter pipe for gas transmission played a significant role in the subsequent larger diameter pipe sizes used by the oil industry.

There has been a massive increase in submarine pipelines in the post-war period. Both oil and gas lines were laid under rivers and in harbours from the earliest days, and the 1920s and 1930s saw many miles of pipeline laid across rivers and in shallow water swamp areas. In the post-war period exploration moved offshore, pipelines followed, and lay barges and reel barges have augmented the earlier techniques of floating or pulling the pipeline into place.

The challenge represented by the discovery of major oil and gas accumulations in ever-increasing depths of water is being met by the pipeline industry and currently it is considered feasible to lay 22-inch line to transport gas in water depths of more than 400 ft, while techniques to lay larger diameter pipe in greater water depths are on the drawing board.

The last quarter of the century has seen a global development in the utilization of natural gas. In the U.S.A. gas pipeline mileage increased more than two-fold to give some 180,000 miles of transmission and approximately 400,000 miles of distribution lines bringing natural gas to all major population centres in the country. The Canadian gas industry, which had its origins in the last century, also saw major expansions in the post-war period, including the Trans-Canada pipeline with a terminal in New York State and the Alberta-

San Francisco line, with a length of some 3750 miles.

The successive discovery and development of gas fields in the Po Valley, Lacq, Groningen, and the North Sea has created the basis for a major natural gas industry in Western Europe. Similar developments in the Soviet bloc place this area second only to North America in terms of transmission mileage. Natural gas has also entered the energy pattern in some less industrially advanced areas, e.g. Pakistan and Bangladesh, while a new facet of the gas industry is the movement of gas across international frontiers – the network of gas lines in Western Europe and the recent completion of the line moving gas from Iran to Russia being examples of this trend.

## PROCESSING

Gas processing is perhaps a less spectacular but nevertheless vital part of the natural gas industry, and its continued evolution has been and remains a critical factor in the economic exploitation of gas reserves.

In the early days in the U.S.A. adequate supplies of sweet (*i.e.* sulphur-free) gas were available; much associated gas (*i.e.* gas produced with oil) was flared, and there was little incentive to treat gas containing sulphur or other undesirable constituents to produce saleable gas. Dehydration of pipeline gas was practised but the main emphasis in processing was the recovery of liquid hydrocarbons (natural gasoline).

The piping of untreated gas produced problems due to the effects of condensation of heavier hydrocarbons and "drips" and "traps" were installed into the early lines to draw off the condensed liquids. With the introduction of the motor car, this condensate changed from a waste product to a valuable commodity and by 1908 the first commercial plant for the separation of casing-head or, as it was later to be known, natural gasoline, was on production. By 1911 well over 100 such plants were operating, mainly in the Appalachian area. These early plants were mainly compression/cooling units, with some expander cooling units also being used, mainly in California. The increasing demand for gasoline created by the First World War gave an additional incentive for improved recovery and this period saw the introduction of the first commercial absorption plants, but patent litigation hindered their general spread until the 1920s.

By about 1923 over 1000 gasoline extraction units were in operation with an average yield less than 1 USG/M cu ft. Absorption plants progressively established a dominant position, with a steady improvement in operating pressures and efficiencies up to the present day.

Apart from the recovery of natural gasoline, early processing was restricted to the reduction of water content. In addition to its corrosive effects, particularly in combination with traces of acid gas, water vapour in contact with gaseous hydrocarbons forms solid hydrates which can exist above its own freezing point, thus presenting a major problem on transmission systems, particularly during cold weather. Early methods of treating the gas employed calcium chloride brine as the drying agent until the late 1930s, when the diethylene glycol and, subsequently, triethylene glycol systems were introduced. Solid desiccants were first introduced in about 1940, the latest form being the molecular sieves which are used extensively for liquefaction units.

The process for the treatment of high-pressure sour natural gas was the Seabord soda ash process introduced in 1920, and by the late 1930s some 50 plants were operating in the U.S.A. and Canada. A number of additional processes suitable for sour natural gas treatment were introduced during the 1930s based on high-pressure scrubbing with a variety of chemicals. The ethanolamines, originally introduced in the form of TEA, progressively established a favoured position for sulphur and carbon dioxide removal from natural gas based on the use of MEA and DEA.

Sulphur recovery from natural gas which had been practised on a limited scale in the 1930s employing, for example the thioarsenate (Thylox) process, was given further impetus during and after the Second World War and the massive sulphur production in both Canada and France (Lacq) in the post-war period by the treatment of sour natural gas affected the world market for this basic chemical. The ethanolamine processes further consolidated their position as the major treatment during the post-war period and were joined by such processes as Benfield, Vetrocoke, Stretford, Sulfinol, etc., which indicated the increasingly international nature of the development gas treatment techniques.

## LIQUEFIED NATURAL GAS

The recent development of the commercial shipment of liquefied natural gas is a most significant factor in the world energy pattern.

Refrigeration was first applied to natural gas in the 1920s for the separation of helium. Pilot schemes for liquefaction and transportation of gas were tried in Hungary in the 1930s, but the first recorded commercial liquefaction plant was built at Cleveland, Ohio in 1940. The plant had a capacity of 4 MM cu ft/day, with LNG storage of about 150 MM cu ft for peak shaving purposes. A major explosion in the storage area did not deter further development, and liquefaction has become an accepted method of storing gas for peak shaving purposes.

In 1951, experiments were carried out to investigate the possibility of transporting liquefied gas from Louisiana to Chicago and by 1956 two barges had been built and tested. The project, as such, was not pursued but, as a result of subsequent development, the *Methane Pioneer,* the first ocean-going LNG tanker, made a Transatlantic crossing with a cargo of LNG for the U.K. in 1959. The success of this project led to the installation of plant to liquefy gas from the Hassi el R'Mel field in Algeria for shipment to the U.K. and France.

The first liquefaction plants used the cascade system, but subsequent units have used the more sophisticated mixed refrigerant process; plant sizes have increased from the 4 MM cu ft/day of the Cleveland plant to 235 MM cu ft/day currently in use in Libya, with even larger plants in the design and construction phase. Storage systems include insulated tankage and frozen underground excavation with insulated roofs.

The first LNG tankers, the *Methane Pioneer* and the *Methane Princess*,• used free-standing tanks of 27,000 cu m capacity. 40,000 cu m tankers are used to move Libyan LNG and LNG is being moved from Alaska to Japan in 71,500 cu m tankers using membrane-type tanks. 125,000 cu m tankers currently on the drawing board are considered by some authorities to be approaching the largest size which can be operated efficiently.

## CONCLUSION

The increasing world demand for energy, the gradual rise in oil prices, and the current emphasis on reduction of atmospheric pollution are combining to increase the demand for new supplies of natural gas throughout the world, and particularly, at this moment, in the U.S.A., where domestic reserves are insufficient to meet consumers' requirements.

Meeting the demand will require greatly increased expenditure for exploration for gas at greater depths and in areas of increasingly adverse operating conditions, and vast capital investment in pipelines and in LNG plants and tankers to move gas to the consumer. The alternative, manufactured gas or SNG, will also require enormous capital investment. Based on this review of the history of the natural gas industry, there can be no doubt that the future technical problems will be overcome, but the investment required can only be justified by realistic prices for gas.

## DISCUSSION

*J. A. Field* (National Coal Board (Exploration) Ltd) said he would like to congratulate his friend Stan Churchfield and his colleagues at Burmah on their remarkably clear and concise review of the technical developments in natural gas discovery, production treatment and transportation.

Their account and Mr Tugendhat's excellent review of the implications of the growth of natural gas provided a perfect background for the Conference.

Until he arrived in the room that morning he had been at a loss to understand why the organizers of the conference should have lighted on him to start this first period of discussion. Looking round the room he realized that the answer probably lay in his white hairs, not for prophecy about the future – that could be safely left to the authors of the later papers on their programme – but perhaps to round off the historical review provided in this first paper. The colour, quantity and quality of the hair on most heads in the room convinced him that his connection with the gas industry was longer than most of the others present.

It was in 1932, 40 years ago, that he first started work in the British
gas industry, in a lowly capacity at the Rotherhithe works of the former
South Metropolitan Gas Company. He wondered perhaps whether there was
any connection in Stan's subconscious between this fact and the reference
in the paper to the first appearance in that year of some creature that he had
described as the "Ubiquitous Bulldozer". Stan and he represented their
respective companies on what was probably one of the largest joint operating
committees dealing with exploration in the North Sea and they had had
their arguments. He admitted that he had tried to get his own way in their
discussions but had not realized that Stan had come to think of him as
ubiquitous or even a bulldozer.

Forty years was a long time, but he wished his audience to note that even
in those far-off days his old company thought in terms of high calorific
value and low sulphur gas. If they could not make 1000 BTU gas, they at
least led the way with a declared calorific value of 560 BTU and at
Rotherhithe they operated a plant for the total removal of organic sulphur.
There were, of course, reasons for this. The most popular story was that
some of the directors still relied on Batswing burners for their domestic
lighting. Those extra BTU made all the difference in the illuminating power
and, of course, without organic sulphur the burners lasted longer.

There had been many changes in those last 40 years, but none so
significant for Britain's gas industry as the arrival of natural gas. Those
there who had been concerned with and about gas over that period would
appreciate the great debt that this country owed to friends and colleagues
in the petroleum industry for the way in which they applied their techniques
and worldwide experience to providing the U.K. with an additional
indigenous source of energy, which had made it possible to bring about such
a revolutionary change in all our thinking about gas and the place that it
should fill in our country's energy economy.

They must pay credit, too, to the gas industry on the way in which it
responded to this enormous challenge and he quoted just a few figures from
the Gas Council's Annual Reports which illustrated this and which they might
bear in mind later in the conference when they heard about the future.
He chose as his datum the year 1964–65. Sales that year for the first time
topped 3000 million therms and the contribution of natural gas started.
The latest figures available were those for the year 1970–71. The figures
quoted were therefore for change over a period of six years.

In 1964–65 19 million tons of coal were carbonized to produce gas;
by 1970–71 this figure had fallen to 3·2 million tons and a total of
4644 million therms or 69 per cent of the total gas available was the
result of natural gas production from the North Sea.

Total sales of gas nearly doubled from 3169 million therms to
6133 million therms, while domestic sales more than doubled from
1723 million therms up to 3653 million.

That it was possible for the gas industry to switch, in such a short period,
from gas manufactured from coal and petroleum products to natural gas
was due to the tremendous effort that the oil industry devoted first to

exploration and then, when natural gas had been discovered, to the
rapid development of production facilities.

The first U.K. production licences were issued in September 1964 and
the first U.K. offshore well was spudded by Amoseas with *Mr Cap* on
Boxing Day, 26 December 1964. Since that date a total of 194 exploration
wells had been drilled on the U.K. shelf in the search for natural gas.
Costs, of course, varied, but a figure around £100 million seemed a reasonable
estimate for the cost of these wells.

A gas discovery was a nice thing to have. Your management was usually
pleased and it could normally be expected to get a headline and five or six
inches in the paper. Indeed, his company's last one got about 15 inches,
probably because it was put out on Sunday for the Monday papers and it
had been a dull weekend. But it was not money in the bank. Indeed, as John
McLean, President of Conoco, told his late chairman, just after Viking had
been discovered, "Alf, all you have got is a licence to spend money."
This was true but it was also the signal for the start of a race to get
production flowing and therefore the start of a really major engineering
exercise, the bringing-together of the team of experts required to design
and to supervise the construction of plant, machinery, and structures
required, the search for contractors with the capability to carry out the
work, and the search for the money that it was going to cost.

Many figures had been quoted for the total investment to date.
As George Orwell might have said, "Some figures are more right than
others." His, if it was wanted, was something between £300 million and
£350 million, in addition to the money spent on exploration. Getting away
from guessing games, it was interesting to note that a total of 145
production wells had been drilled and that the gas from these wells was
flowing through 153 miles of undersea pipeline. That last figure would be
increased to 236 miles when the Conoco/NCB Viking field came on stream
that year.

This part of the business seldom got into the headlines and he suspected
that few people outside the oil industry really appreciated the enormous
effort and investment that had gone into the development of natural gas
from the five major discoveries in the U.K. North Sea and, as a result of
which, the HiSpeed gas revolution had been possible

Having been associated closely with one of these discoveries, he was very
happy to have an opportunity to congratulate all those concerned in the
oil industry on the way in which they had helped to bring this about.

# The Future Climate
# for the Exploration for Gas

## By A. F. FOX

### (British Petroleum Co. Ltd)

## SUMMARY

The level of exploration activity, whether it be for oil or gas, ultimately depends upon the two factors of demand and economic return. In this paper, the future demand is estimated and discovery costs are examined very generally. The need for the discovery of additional gas reserves is examined and it is concluded that additional discoveries will probably be incidental to exploration for oil.

The development costs of an offshore field are examined and the development in concession terms described. The effects of various methods adopted by governments to obtain a share in the return from successful operations are illustrated by discounted cash flow examples and some conclusions upon economic pricing drawn.

The future prospects for exploration directed to the primary object of discovering gas rather than liquid hydrocarbons will be controlled mainly by the future demand for energy in this form. This demand in its turn will depend upon the relative cost, including royalties and taxation, of hydrocarbon gas and competing fuels and the overriding consideration of the existence of accumulations of gas which can be found with techniques of exploration available and their location.

## THE GROWTH OF DEMAND

Before looking ahead, it is necessary to look backwards also to appreciate today's situation. The rate at which the world production of oil and gas has been increasing in consequence of the rising demand up-to-date is shown on the diagram (Fig. 1). The data on which these curves are based are taken from U.S. Bureau of Mines publications for oil production and the U.N. statistical yearbooks for gas production. The original data for gas are not consistent, the basis of reporting having been changed in 1955, and a correction factor has been applied to the pre-1955 figures to make them appear comparable with later data; because of this, any extrapolation must be treated with some reserve. All quantities in this paper are given in therms and are rounded off, although the original data appeared in various units. If nothing occurs to interrupt the steady increase in oil and gas production, it can be conservatively expected that annual world production of oil in 1980 will be about $1400 \times 10^9$ therms, possibly rather higher, whilst projection to the year 2000 becomes too hazardous to attempt. The production of gas can be expected to reach $650 \times 10^9$ therms by 1980 and

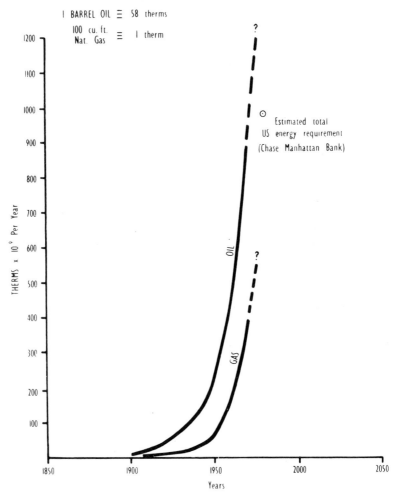

Fig. 1. World oil and gas production. Therms x $10^9$. Data adapted from U.S. Bureau of Mines in World Oil, 15 August 1955 and 15 February 1969. U.N. Statistical Yearbooks, 1959, 1967 and 1970. Erdol and Kohle. February 1961.

perhaps $1250 \times 10^9$ therms by the year 2000 A.D. These extrapolations must be regarded with extreme caution, since it is most unlikely for the economic considerations which have been the reason for the historical increases in hydrocarbon production to continue unchanged for more than a decade to come. However, unless there is a catastrophic revaluation in the world's requirement for and use of hydrocarbon fuels, we may expect the demand for both oil and gas to increase very considerably over the next decade or so.

It is interesting to note that the part of world energy supplied by hydrocarbon liquids and gases has not been divided between the two forms in the same proportion over the years. In Fig. 2, a curve has been constructed to

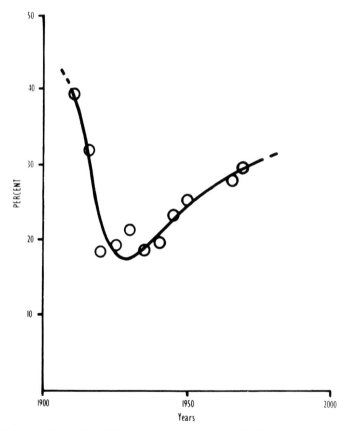

Fig. 2. Proportion of world petroleum energy provided by petroleum gas. Based on production data from U.S. Bureau of Mines and U.N. Statistical Yearbooks.

show the proportion of the total energy provided by petroleum which has been supplied by gas. It can be seen that, after 1900, the energy supplied by gas dropped from over 33 per cent in 1910 to a low of 18 per cent in 1930. Since that date, gas has steadily regained its position and, in 1970, supplied almost 27 per cent of the total.

We have no basis for extrapolating the curve but it is thought that it will level off somewhere between 30 and 40 per cent.

## GAS UTILIZATION

The picture for industrial and commercial consumption, however, is quite different. The indications are that the U.K. is entering a period of rapidly increasing industrial use of petroleum gas due to a number of very large industrial users adopting gas as their primary fuel. How long or how far this process will go will depend upon both the continuance of an adequate supply and the maintenance of a competitive price. It would be naive to extrapolate the graph of total gas consumption too far. Indeed, the reserves of gas so far discovered in the North Sea are believed to be equivalent to between 303 and $405 \times 10^9$ therms and to be capable of sustaining a production of the order of $40 \times 10^6$ therms a day or $14.6 \times 10^9$ therms a year. Therefore, unless additional major discoveries are made, the total consumption cannot increase much beyond this figure without substantial imports and, since the price ruling in the U.K. is rather lower than that in the rest of Europe, this is unlikely. With the discovery of large oil reserves in the North Sea, a certain amount of associated gas will become available. At the time of writing, it is not known what gas/oil ratios will be found or what proportion of gas may be available. It is considered, however, that the net reserves which might result from the presently announced discoveries could be about $202 \times 10^6$ therms, the additional daily production from which would permit consumption to increase to $60 \times 10^6$ therms a day.

The largest demands for gas are likely to arise wherever there is a concentration of industry. It is true that centres of population, without necessarily having adjacent industry, do generate a demand but, even in the U.K., where the use of gas for domestic purposes is well established, a reasonable extrapolation of the curves in Fig. 3, which is based upon the Gas Council's report for 1970, shows that the average annual consumption of gas by each household in the U.K. can be expected to rise only to about $0.3 \times 10^3$ therms a year, with a total domestic consumption reaching a maximum of $4.0 \times 10^9$ therms a year and probably not increasing much beyond that. Domestic usage may therefore eventually be about 38 per cent of the industrial and commercial usage.

It is instructive to compare the demand patterns in the U.K. and the U.S.A. from which data are available. First, in the U.K. the pattern was developed in the days of town gas produced from coal; the source of supply has altered in the last decade from the position in 1960, when over 80 per cent

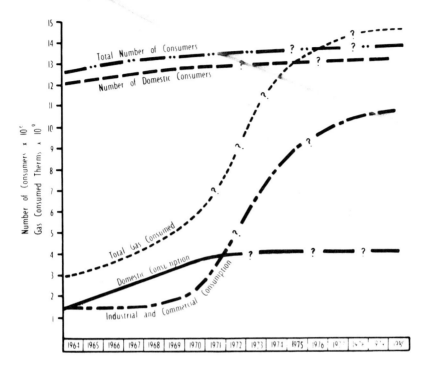

Fig. 3. U.K. gas consumption. Derived from figures published in the Gas Council Report for 1970.

of gas was produced from coal, the remainder being produced from petroleum; through the situation in 1968, when coal provided under 20 per cent, petroleum-derived gas about 55 per cent and the remainder was natural petroleum gas; to the position today, where natural gas provides some 70 per cent of the total, gas produced from petroleum just over 20 per cent, and only somewhat under 10 per cent is still produced from coal. Indeed, the town gas which is produced from coal arises as a by-product of steel coke-ovens. Furthermore, in the U.K., town gas was the first easily distributable form of energy available whilst, being manufactured from coal, its price was necessarily higher per therm than coal itself, which normally meant that gas did not compete with coal for the large-scale industrial market. The early developments were therefore for domestic and industrial lighting and for cooking, with some development for space heating. The advent of cheap petroleum gas in quantity therefore found a distribution network already established and a habit of gas usage well developed.

In the U.S.A., natural gas satisfies 31 per cent of the energy market and is second only to oil, which provided 44 per cent. The Chase Manhattan Bank's publication "Outlook for Energy in the United States", published in 1968, from which the above figures were drawn, also divides the 1968 energy market into four:

| | |
|---|---|
| Industrial and commercial | 39% |
| Transport | 24% |
| Electric utilities | 22% |
| Residential | 15% |

and estimates that the total energy consumption will increase by $1.2 \times 10^9$ therms per day between 1965 and 1980, during which time the consumption of gas will increase from some $450 \times 10^6$ therms/day to $659 \times 10^6$ therms/day.

Although the pattern of development of the supply of energy has been quite different in the U.S.A. from what it was in the U.K., it is interesting to note the similarity of the proportion of domestically-used energy (i.e. the "residential" classification) to industrially-used energy. If all forms of energy are considered, it is 36 per cent and if gas only is considered it becomes 37 per cent. These figures are almost identical with the 38 per cent obtained above for the U.K.

The major areas of demand can therefore be definitely established as the main centres of industry and these in the foreseeable future, as at present, are in Western Europe, the U.S.A. (particularly the eastern states), and Japan. Lesser demands would arise in areas with large populations, such as SE Australia and parts of India, but are not likely to do more than stimulate local exploration.


## DISCOVERY COSTS

As was mentioned at the beginning of this paper, the second control on the development of gas will be an economic one. In the U.S.A., new reserves of petroleum discovered for each dollar spent upon exploration have remained at a fairly constant figure since 1955 at about 87 therms (1.5 brl of oil equivalent) per dollar, whilst gas has formed about 50 per cent of the reserves found over the same period. There is no reason to suppose that this cost will decrease in the next decade and every reason to believe that it must increase as the available reserves are progressively used up. It is also probable that the ratio of reserves to production will move still further downward and (in the U.S. at least) a reserve of ten times the annual production may be difficult to maintain. This ratio for the U.S.A. cannot be directly compared with similar ratios derived for other areas since the basis of reporting reserve discoveries is fundamentally different.

It has been estimated that the use of gas in the U.S.A. will reach $659 \times 10^6$ therms/day ($11.5 \times 10^6$ b.o.e./day) in 1980 and that this will be 25 per cent of the total energy demand at that date ($2.6 \times 10^9$ therms/day).* The figure of 25 per cent of total use appears rather lower than the world average from Fig 2, which looks to be nearer 30 per cent. However, averages in individual countries will naturally vary from the overall figure. The average consumption over the 11 years for 1970-1980 inclusive is therefore estimated to be $550 \times 10^6$ therms/day, which will represent a total consumption of $2.2 \times 10^{12}$ therms. The gas reserves of the U.S.A. were estimated to be $2.90 \times 10^{12}$ therms in 1969 and these reserves, less the production 1970 - 1980 inclusive plus the reserves discovered, must equal the reserves required in 1980 which at a ten-year reserve bank would be $2.4 \times 10^{12}$ therms and therefore:

$$2.90 \times 10^{12} - 2.2 \times 10^{12} + R \qquad = 2.4 \times 10^{12} \text{ therms}$$
$$\text{or}$$
$$R \qquad = 1.7 \times 10^{12} \text{ therms}$$

If in the U.S.A. the gas equivalent of $1.7 \times 10^{12}$ therms can be found, the exploration cost incurred will be greater than $\$20 \times 10^9$ and, if this expenditure is equally spread over the decade and financed from sales of gas, the charge will be about just under one cent per therm.

It is not possible to reproduce the calculation above for the U.K. or even for Western Europe, since the basic data are not available. Nevertheless, some pretence at estimating can be made using estimates of the volume of reserves so far discovered and of the costs of exploration.

At the end of 1971, the number of wells drilled in the U.K. sector of the North Sea was 371, of which 231 were classed as exploration or appraisal wells, and 140 as development wells. All the development wells were drilled in the southern part of the North Sea and may be considered to have been drilled with the primary objective of producing reserves of gas. Forty of the exploration and appraisal wells were drilled in the northern waters, and these wells may be considered as being located with the principal objective of finding liquid oil and associated gas.

It is impossible to estimate the cost of this drilling with great accuracy but it is probable that an average figure of £800,000/well, inclusive of geophysical and overhead costs, would give a total not too far from the truth, the southern wells averaging somewhat less and the northern considerably more. This would mean a total expenditure of about £185 million on exploration, divided into £32 million on oil exploration and £153 million on exploration for gas. It is probable that in cash terms the division should be weighted in favour of gas, since wells in the northern areas are much more expensive; possibly £40 million for oil and £145 million for gas would more accurately represent the split.

There were 382 exploration and appraisal wells drilled in the whole of the

* "Outlook for Energy in the United States". Chase Manhattan Bank N.A., New York, October 1968.

North Sea up to the end of 1971, of which 262 in the south can be considered as drilled primarily to locate gas and 120 located primarily for oil. That is an estimated expenditure of £306 million which may be arbitrarily divided between gas £200 million and oil £106 million.

In return for these huge expenditures, the estimated reserves discovered are possibly of the order of:

| Oil | North Sea | $520 \times 10^9$ therms |
|---|---|---|
| | U.K. zone only | $350 \times 10^9$ therms |
| Associated gas | North Sea | $210 \times 10^9$ therms |
| | U.K. zone only | $45 \times 10^9$ therms |
| Unassociated gas | North Sea | $330 \times 10^9$ therms |
| | U.K. zone only | $280 \times 10^9$ therms |

These figures must be treated with considerable reserve since they are based on very imprecise data and possibly unreliable press reports. Moreover, they are not comparable with the published figures for the U.S.A., which are arrived at by a summation of returns made by companies under statutory requirements generally based on production capacity. The figures deduced here are guesses based on inferred reservoir sizes and estimated rock properties and do not take into account the expected final extent of every announced discovery.

Costs per therm may therefore be calculated and, using the estimates given above, the total petroleum energy found in the North Sea ($1060 \times 10^9$ therms) has cost £306 ($800) million and some 1325 therms have been found for each dollar expended upon exploration in the whole of the North Sea, 15 times more than the United States figure. If unassociated gas only is considered, the figure becomes 630 therms/dollar expended, the total of $330 \times 10^9$ therms having cost £200 ($524) million.

These figures are apparently a good deal more attractive than the expectation for the U.S.A. and although the difference in the meaning of the "reserves" given for two areas makes any direct comparison impossible the difference of an order of magnitude cannot be ignored.

## THE DISCOVERIES REQUIRED

Although the costs of new discoveries in the North Sea may be less than in the U.S.A., no data are available at present to provide equivalent figures for other areas. However, the criterion which would decide whether exploration for oil and gas should proceed in any large unexplored sedimentary basins would be that discovery volumes should be at least of the same order as those for the North Sea. Even though exploration costs in areas very remote from centres of industry turned out to be considerably higher, the therms found per dollar expended should still be over 500 in order to make the adventure worthwhile.

The future demand, too, will continue to rise, it is thought, more or less on the same lines as heretofore, although this rise may be modified by the price which has to be paid by the consumer. We examine next the quantities which may be required.

The cost of discovery outside the U.S.A. and in new offshore basins, in which it is considered future developments will be concentrated, is expected to approximate to that for the North Sea or to be rather higher, *i.e.* to be in the region of 1000 therms per dollar expended. If we then assume that the growth in world consumption of petroleum will increase without break (*i.e.* along the upper lines of the graph in Fig. 1), the total consumption will be of the order of $1400 \times 10^9$ therms per year by 1980. In 1980, from Fig. 2, it can be seen that petroleum gas should form about 32 per cent or a little less of this figure, *i.e.* about $420 \times 10^9$ therms.

If we assume that a ten-year reserve bank is sufficient, the reserves necessary to maintain this figure will be $4.2 \times 10^{12}$ therms. However, although a ten-year reserve bank may be possible when using figures quoted for the U.S.A. reserves, when considering worldwide figures a 20-year reserve bank is more reasonable and this figure is $8.4 \times 10^{12}$ therms. Gas reserves in 1969 were estimated to be $13.85 \times 10^{12}$ therms, whilst the production 1970 - 80 inclusive is estimated to be $4640 \times 10^9$ therms and, therefore, the reserves to be discovered (R) can be calculated as follows:

For a ten-year reserve bank:

$$13.85 \times 10^{12} - 4.64 \times 10^{12} + R = 4.20 \times 10^{12}$$
$$R = -5.01 \times 10^{12} \text{ therms}$$

For a 20-year reserve bank:

$$13.85 \times 10^{12} - 4.64 \times 10^{12} + R = 8.40 \times 10^{12}$$
$$R = -0.81 \times 10^{12} \text{ therms}$$

which indicates that the worldwide reserves of gas known today are sufficient to sustain production up to 1980 without further discovery. However, this conclusion does not take into consideration the location of the gas reserves or the relative cost of getting usable gas to market. Also, after 1980, additional reserves will certainly be required and, as the exploration for, and discovery and development of, a gas field probably takes ten years to complete, there can be no let-up in the present exploration effort.

It is doubtful whether any new exploration venture starts out with the primary objective of discovering gas, although some (the Southern North Sea venture, for example) started with the knowledge that gas discoveries were much more probable than discoveries of oil. In considering where additional gas reserves may be found, the overriding factor will almost certainly be the discovery of oil incidentally producing additional reserves of associated gas or the discovery that a particular basin does, in fact, yield gas rather than oil. This latter situation appears to be the case in the West Australian basin, where the discoveries so far announced have all been of dry gas although it is difficult to imagine that the venture in such a remote and inhospitable area was embarked upon without the anticipation of an oil discovery.

## DEVELOPMENT COSTS

The cost of the development of a gas field cannot reasonably be estimated until its location and, particularly if it is an offshore discovery, the distance from shore, the depth of water, and the local weather conditions are known. The development cost will also be influenced to a considerable extent by the depth and type of the reservoir rock, by its physical characteristics, by the production mechanism which can be developed, and, of course, by the size of

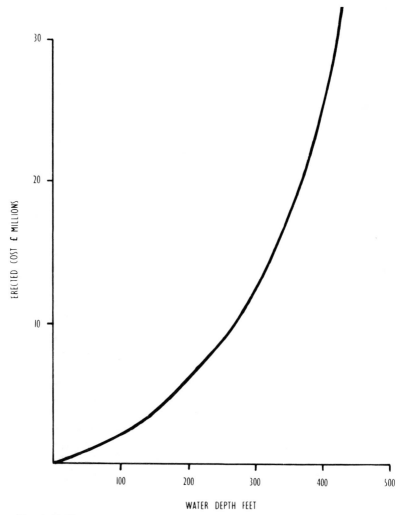

Fig. 4. Drilling and production platforms. Erected cost versus water depth.

the field. The location of the field will impose certain inescapable factors in the
location of wells, treatment plants, and pipelines. These problems become more
acute in offshore fields, where costs are much greater, and development will
normally be from multi-well platforms from which wells will be deviated to
drain as great an area as possible, limited only by the deviation angle which can
be achieved. Multi-well platforms are not suitable for shallow fields because
sufficient deviation cannot be obtained and there is an economic limit to the
maximum depth of water for which they can be constructed, since the cost of
them increases very rapidly with depth of water (and also with the surface area
of the platform). Fig. 4 illustrates the range of cost with depth of water of typical
North Sea platforms. There is unlikely to be a great difference in the cost of
a platform constructed to house gas wells and gas production equipment and
one designed for oil wells and oil production equipment.

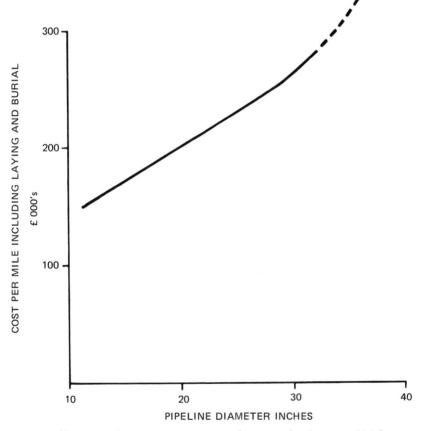

Fig. 5. Offshore pipeline construction costs for water depths up to 200 feet.

The cost of an underwater pipeline, too, increases both with distance from shore and depth of water. Fig. 5 shows the cost per mile for different sizes of offshore pipelines in water depths of up to 200 ft (30 fathoms). It is known that costs for laying lines in depths of up to 450 ft will be much greater than this. However, the oil industry has little experience of pipe-laying in these depth and cost estimates will become more reliable as experience is obtained.

Drilling costs can be estimated with reasonable accuracy once the depth of wells is known, and Fig. 6 shows typical North Sea drilling costs. These costs, however, will vary for different operators, since the figures quoted are "all-in" costs and include all the items charged to the well. Different operators may

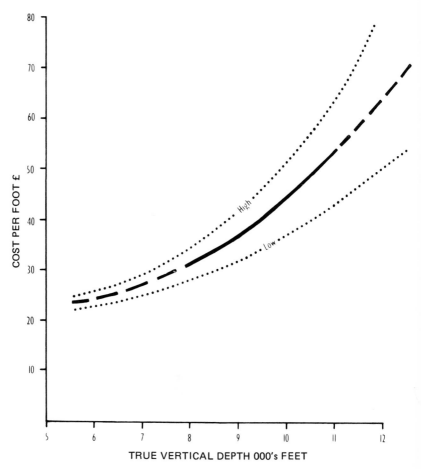

Fig. 6. Estimated drilling costs deviated offshore platform development wells.

consider different items to be chargeable and will, therefore, have different cost figures.

The number of wells will depend upon their productivity and ability to drain a finite area of the reservoir in a reasonable time. These factors are associated with the size of the field, the permeability of the reservoir, and the efficiency of the drive mechanism. Other factors will also enter into the calculation and, of course, the rate of depletion of the reservoir will be considered.

A reasonable size of offshore gas field might yield $5.0 \times 10^{12}$ standard cu ft of gas. If it is assumed that 70 per cent of the gas underground were recoverable, the depth of the reservoir about 10,000 ft, and the reservoir pressure 4500 psi, the gas volume in the reservoir would be $27.5 \times 10^9$ cu ft. Making the further assumptions of 20 per cent rock porosity and 20 per cent interstitial water content, this volume of gas would occupy a rock volume of $172.0 \times 10^9$ cu ft and, if the average thickness of the gas column were 200 ft, the productive area would be $860 \times 10^6$ sq ft or 31 sq miles.

At a reservoir depth of 10,000 ft, the maximum economic distance a well could be deviated is also about 10,000 ft and, therefore, the drainage area for a single platform would be about 11 sq miles, indicating that three platforms would be required to drain the whole area of the reservoir and, if these are constructed in 100 ft of water, they would cost £6 million (Fig 4).

The number of wells required depends upon individual productivity, but if 14 wells were drilled from each platform, a total of 42 wells would probably represent between 30 and 35 wells productive at any one time and an average well productivity of about 20 mmcf/day. The cost of 42 deviated wells to 10,000 ft would be £18.9 million, say, £19 million (Fig 6).

The cost of a 30-inch pipeline 100 miles long would be £26 million.

To these amounts, the further costs for the production, processing, and control equipment on shore and on the platforms must be added, say, £6.0 million, making a total cost of £57 million, say, £60 million in all.

## CONCESSIONS

There has been a steady stiffening in the terms granted by governments since the end of the 1939-45 war. This has come about for three main reasons. First, the lessening of exploration risk in many areas, for instance in the Middle East during the years when preliminary exploration was being carried out without knowledge of the prospects, concessions were awarded over large blocks of territory, indeed over whole countries, whilst today, where the province is a proven and prolific oil-producing area, concessions are more limited in area and scope and more demanding in financial terms. Secondly, the presentation of a common front by the Organisation of Petroleum Exporting Countries has enabled them to stiffen their terms and, thirdly, the increasing sophistication of the governments awarding concessions who can now see, from recent past history, the effects of successful and unsuccessful exploration and are able to adjust the legal framework within which they will allow the exploration for, and development of, oil and gas reserves to take place to the best advantage of their own country.

There has also been an increase in the degree of competition in the international business caused by the setting up of national companies both by consuming and producing countries, by the increasing activity of the so-called "independent" companies, who are often able to fund overseas exploration ventures with the aid of favourable taxation provisions in their home countries, and to some extent by the activities of individual entrepreneurs, who are able to obtain and subsequently to farm out the smaller concession areas now usual.

There has also been a marked increase in tempo visible, particularly over the past decade. This may be the result of the greater demand for oil but it is certainly increased by the short time-limits now put upon the evaluation of new territory, coupled with the fact that, once a reservoir is discovered, economic forces compel its development, provided the development can be financed profitably even if the exploration expenses are not recovered.

For simplicity in discussion we may consider three broad types of concession.

1. The large area long-term concession
2. Limited area/term licence with relinquishment
3. Contract operation

The first type was the favoured type in areas where the risk was high, exploration expenses likely to be large, and returns problematical. Many of them were awarded in the 1930s and 1950s for terms of 50-75 years. All have now been amended to include provision for relinquishment of parts of the original area as exploration progresses. They usually called for an initial cash bonus on signature and now have a government tax rate of 55 per cent of the notional profit calculated from posted prices and on a royalty of 12½ per cent expensed, although the actual detailed payments may fall under several heads. This type of concession appealed only to a major integrated oil company because the resources required were so large that only an organization with virtually unlimited capital resources could hope to pay the initial bonus and to maintain the logistic effort necessary to carry out adequate exploration. The effect was to identify the company closely with the country in which it expected to continue working for a generation or more. Development could be and was planned and carried out with very long-term objectives and it is probably not chance that the science of reservoir control was developed by a company enjoying this type of concession.

Nevertheless, this type of concession has now fallen into disuse, although many still continue in amended form. The more rapid results obtained by the other forms of agreement have outweighed, in the opinion of the governments, the more conservative virtues associated with the older form. It cannot be denied that a more rapid cash flow can be obtained by other methods, it is at least open to argument whether more economical overall development can be so obtained. For an undeveloped country the slower initial pace of development may be a great advantage in that it allows time for the training of nationals in the special skills required by the industry and also for a build-up of the industrial life of the country so as to enable it to participate in the economic advantages associated with a successful oil venture.

The limited area/term licence is typified by those now being awarded in the North Sea by the U.K., the Netherlands, and by Norway. The areas covered by this type of licence vary somewhat from place to place and are usually within the range 100-500 sq miles. The length of time allowed is normally 20 years or more (*i.e.* the minimum time to drain reservoirs by methods accepted as reasonable practice) but the pace of exploration is enforced by setting a short term of years (normally three or five) after which a proportion (normally one half) of the area must be relinquished. This type of concession allows much greater competition, ensures an intense exploration effort in the first five years or so after award, and, if discoveries are made, ensures the rapid development of a producing industry. However, because of the immediate build-up of effort — and very often the equally rapid decline in interest — this form of concession encourages an itinerant exploration industry which does not develop roots in the host country and is not in any way identified with it, very often not even employing any nationals within its ranks. Indeed, it is possible to envisage an offshore concession of this type completely explored by geophysical parties and drilling outfits which never set foot upon the shores of the host country, are manned entirely by ex-patriate staff, and the results of whose work are evaluated on another continent.

The contract type of agreement is a variant of either of the above types, the essential difference being confined to the national ownership of any crude oil found.

The differences in these types of concession are overshadowed by the variations in the methods of taxing the profits of the companies. The methods adopted are of three main types:
1. Capital lump sum payments
2. Taxation upon realized income
3. Participation in production

The first type form the signature bonuses and bonuses paid on reaching a set production figure, forced building of schools, hospitals, ports, etc. The second type comprise taxes of all kinds, royalty payments, licence fees, rentals, training levies, etc. The third form involves the government in taking a direct share in the production effort, with an equivalent share of the product which may be sold back to the operating company. Much can be argued for and against the various combinations which have been adopted from time to time but ultimately the only true criterion is the effect upon the cash flow of the operating company. Three examples to illustrate this are given below.

## THE EFFECT OF CONCESSION PAYMENTS

Four separate cases have been considered, all of them in terms of cash flow before tax. The first and simplest case represents in simple terms the old type of concession given on terms which involved the exploring agent in no payments except royalty and taxes on income. This type of concession meant that the size of concession taken was fixed only on the agent's resources and, for the purpose of this exercise, this was fixed at £150 million (in fact, the Discounted Cash Flow cumulative net outlay came out to £149 million). It has been assumed that ten units were taken up, the exploration cost for each of which was £15 million spread equally over three years. It has also been assumed that one producing field was found and developed at a total cost of £60 million. The cash flow, all discounts being taken at 15 per cent, is shown in Table I.

In this case, the cumulative net outlay would be £190 million with a present worth of £149 million and the present worth of the adventure + £23 million. This type of arrangement is, of course, ideal for highly speculative acreage, since the actual investment in exploration can be maintained if there is encouragement or the concession can be relinquished if results do not warrant continued exploration. It is usually modified by the introduction of a production bonus payable when production reaches a certain figure. Table II illustrates the effect on the cash flow of a production bonus of £50 million in year 6, which reduced the present worth to zero.

### TABLE I
### Cash Flow — No Concession Charges
£ millions

| Year | Exploration | Development | Income | Cash flow | P.W. @ 15% |
|------|-------------|-------------|--------|-----------|------------|
| 0    |             |             |        |           |            |
| 1    | -50         |             |        | -50       | - 47       |
| 2    | -50         |             |        | -50       | - 41       |
| 3    | -50         | -20         |        | -70       | - 49       |
| 4    |             | -20         |        | -20       | - 12       |
| 5    |             | -20         | +50    | +30       | +16        |
| 6    |             |             | +50    | +50       | +23        |
| 7    |             |             | +50    | +50       | +20        |
| 8    |             |             | +50    | +50       | +18        |
| 9    |             |             | +50    | +50       | +15        |
| 10-20|             |             | +50    | +50       | +80        |

TABLE II
**Cash Flow – Production Bonus Only**
£ million

| Year | Bonus | Exploration | Development | Income | Cash Flow | P.W. @ 15% |
|---|---|---|---|---|---|---|
| 0 | | | | | | |
| 1 | | -50 | | | -50 | -47 |
| 2 | | -50 | | | -50 | -41 |
| 3 | | -50 | -20 | | -70 | -49 |
| 4 | | | -20 | | -70 | -12 |
| 5 | | | -20 | +50 | +30 | +16 |
| 6 | -50 | | | +50 | zero | zero |
| 7 | | | | +50 | +50 | +20 |
| 8 | | | | +50 | +50 | +18 |
| 9 | | | | +50 | +50 | +15 |
| 10-20 | | | | +50 | +50 | +80 |

The present worth of the production bonus is £23 million.

The third case considered is one where a rental is paid. It has been assumed that five units were originally taken up, each unit having a rental of £2 million/year. After three years exploration, half the area was given up and one field was developed. Table III illustrates this position.

TABLE III
**Cash Flow – Rental Basis**
£ millions

| Year | Rent | Exploration | Development | Income | Cash Flow | P.W. @ 15% |
|---|---|---|---|---|---|---|
| 0 | | | | | | |
| 1 | -10 | -25 | | | -35 | -33 |
| 2 | -10 | -25 | | | -35 | -28 |
| 3 | -10 | -25 | -20 | | -55 | -39 |
| 4 | - 5 | | -20 | | -25 | -15 |
| 5 | - 5 | | -20 | +50 | +25 | +13 |
| 6 | - 5 | | | +50 | +45 | +21 |
| 7 | - 5 | | | +50 | +45 | +18 |
| 8 | - 5 | | | +50 | +45 | +16 |
| 9 | - 5 | | | +50 | +45 | +14 |
| 10-20 | - 5 | | | +50 | +45 | +72 |

The present worth of the investment here is £115 million, the present worth of the adventure £39 million, whilst the government concerned obtained a total take of £115 million with a present worth of £46 million. This sort of arrangement can also carry a production bonus and if a £50 million production bonus were payable in year 6 as before (present worth £23 million) the present worth of the adventure drops to £16 million, whilst the P.W. of the government take rises to £69 million.

The final case illustrated is that of the initial or signature bonus, which may, of course, be negotiated or obtained by auction. In this case, it is assumed that five units were again taken at a total bonus cost of £50 million. They were explored and one field developed at the same costs as have been used before. Table IV illustrates the position where the present worth of the total outlay is £137 million, the government take £50 million, both actual and present worth, whilst the present worth of the adventure is + £35 million.

### TABLE IV
### Cash Flow — Signature Bonus or Auction Fee
£ millions

| Year | Bonus | Exploration | Development | Income | Cash Flow | P.W. @ 15% |
|------|-------|-------------|-------------|--------|-----------|------------|
| 0 | -50 | | | | -50 | -50 |
| 1 | | -25 | | | -25 | -23 |
| 2 | | -25 | | | -25 | -20 |
| 3 | | -25 | -20 | | -45 | -32 |
| 4 | | | -20 | | -20 | -12 |
| 5 | | | -20 | +50 | +30 | +16 |
| 6 | | | | +50 | +50 | +23 |
| 7 | | | | +50 | +50 | +20 |
| 8 | | | | +50 | +50 | +18 |
| 9 | | | | +50 | +50 | +15 |
| 10-20 | | | | +50 | +50 | +80 |

A £50 million production bonus in year 6 would reduce the present worth to £12 million and increase the present worth of the government take to £73 million.

The important factors in these several cases are summarized in Table V below:

TABLE V

**Cash Flow Summary (Discounted at 15 per cent)**

| | P.W. outlay | P.W. adventure | Government take | | Area explored |
| | | | Gross | P.W. | |
| | £ million | £ million | £ million | | Units |
| Case 1 | 149 | +23 | zero | zero | 10 |
| Case 2 | 149 | zero | 50 | 23 | 5 |
| Case 3 | 115 | +39 | 115 | 46 | 5 |
| Case 4 | 137 | +35 | 50 | 50 | 5 |

Whilst no argument as to the most satisfactory method of collecting the government share can be deduced from this, it is interesting to see that the case in which the government take is restricted to income taxes seems to result in the lowest present worth for the company concerned because of the large exploration effort made. The initial bonus or auction method appears to result in the highest present worth for the government but the rental method, whilst giving the company the highest present worth, gives the government a much larger gross take. In this connection it should be noted that the discount rate for the government share, be it signature bonus or auction fee, production bonus, rental, royalty, or tax, would never be as high as the commercial rate appropriate to a trading concern. The market rate for money would probably be the appropriate rate to take and this would improve cases 2 and 3 in the government's favour.

Much ill-informed advice has recently been published, most of it claiming to show that the auction method is superior in increasing the "taxpayer's" share of the profits derived from the discovery and development of oil and gas. Whilst there seems no good reason why the "taxpayer" should obtain more favourable treatment from the developers of oil than from anyone else, the illustrations given above should serve to show that such facile assumptions cannot be substantiated in practice. The actual method adopted should be sufficiently flexible to maintain exploration interest, to ensure an even level of exploration effort lasting over as long a period of time as is possible, and to obtain for the country concerned a reasonable return of money capital in exchange for the lost mineral asset. It should also be such that the adventurers can obtain, and can see and foresee that they can obtain, a return commensurate with the risks they take and the proportion of success attained world-wide, so that a reasonable proportion of future ventures may be financed from the returns from existing and successful developments.

## GAS PRICE

The hypothetical gas reservoir considered in the previous sections was postulated as having recoverable reserves of $5.0 \times 10^{12}$ cu ft or $50 \times 10^9$

therms, whilst the gross revenue was estimated at £50 million/year for 15 years. If we assume that 75 per cent of the reserves are produced in these 15 years, the average annual rate of production would be $2.5 \times 10^9$ therms and a net realized price of 2.0p/therm would provide the required revenue. Operating costs and royalties probably amount to about 0.3 or 0.5p/therm and, therefore the average gross realized price should be about 2.3 or 2.5p/therm.

Additional operating costs might be added in cases where fields are remote from markets and transport cost must be covered by gross realized price. These costs depend directly on distance and they are therefore omitted.

It is interesting to speculate upon the actual realized price under the current Gas Council contracts but it is difficult to imagine that any of the present operators are able to realize more than 1.25p/therm. At this price it would be impossible to obtain a positive present net worth if any exploration costs were included in the calculation and quite out of the question to generate enough capital to support further exploration. Had the price realized today been known before exploration started, it is unlikely that any company could have carried out any exploration for gas at all and it now seems most unlikely that further exploration for gas will take place in U.K. waters. Increases in gas reserves may, however, result from further discoveries of oil with associated gas.

The cost of possible alternative fuels is also relevant to the general price question. Naturally, the actual alternative in any particular area depends upon the other fuels available there, but in general we may consider three, standard kerosine, heating gas oil for domestic consumption, and fuel oil for electricity generation. Representative average costs, inclusive of duty per therm in the U.K. of these fuels in 1968 were:

| | |
|---|---|
| Standard kerosine | 5.82p/therm* |
| Heating gas oil | 5.16p/therm* |
| Oil fuel | 2.32p/therm† |

    (for electricity generation)

The margin over the realized price of gas is immediately evident.

It is of course, difficult to provide any representative figure for world markets but it may be worth noticing that the published import prices for crude oil for 1971 for the U.S.A. and Japan were:

U.S.A. average duty paid import price December 1971 − 1.84p/therm††
Japan average duty paid import price July/Sept 1972 −  1.66p/therm†††

It is probable that prices in 1976 will show an increase of about 20 per cent over these figures to take account of the  Teheran and Geneva agreements, the

*     Derived from SM & BP Scheduled Prices
†    DTI Digest of Energy Statistics (1971 edition)
††   Cif price from *Platt's Oilgram*
†††  Cif price from Japanese Ministry of International Trade and Industry figure

estimated effect of current short haul negotiations, forecast freight costs, and the prospective increase in demand for low sulphur crude. It has not been practicable to derive product prices from these figures but the landed price of LPG or gas must be governed by them and the operating margin appears to be very small.

## ACKNOWLEDGEMENTS

Whilst the opinions and illustrations used in this paper are my own, I must acknowledge the considerable help I have received from other members of BP's staff. In particular, I would like to thank Miss L. O. Scott and Miss P. E. Bimpson of the Geological Information Branch for unearthing most of the data M. A. Armstrong for providing information on concessions, E. K. Westcott for carrying out the discount calculations, and Miss P. J. Norris for reading and correcting the draft. I would also acknowledge the invaluable help given by T. Fothergill, G. A. Hogg, and W. J. Saint in discussions.

The paper is presented with permission of the Management of The British Petroleum Co. Ltd.

## DISCUSSION

*E. V. Corps* (De Golyer and MacNaughton) said it gave him much pleasure to open the discussion on Tony Fox's excellent paper on "The Future Climate for the Exploration for Gas". Day-to-day weather forecasting was a hazardous occupation, and long-range even more so. The forecaster never satisfied anybody, and Mr Fox knew that his paper would find many critics. There were methods which they thought could be better, basic data whose details were not those they would have chosen, the deductions were not made in the way they would have made them. But these were matters of detail, and had not affected the paper as a whole. It was clear, it was logical, and its message, though possibly somewhat muted, was sound. He was not going to pick holes in the details but would try to fill out a few areas where the subject matter of other papers impinged on Mr Fox's subject, and to draw attention to one important change in the basic data.

The section dealing with the future growth of U.S. gas requirements utilized the 1968 Chase Manhattan forecasts. These had been superseded by events in the three years or more since they were prepared, and results showed that they were erroneous to a significant degree. They forecasted U.S. consumption rising to $659 \times 10^6$ therms/day in 1980. Demand in 1971 was known to have been $670 \times 10^6$ therms/day, with actual consumption $650 \times 10^6$ therms/day. There was therefore a deficit in demand, unfulfilled even allowing for gas from all sources including imports, amounting to about $20 \times 10^6$ therms/day. The Federal Commission's latest report (Feb 1972) estimated that by 1980 demand will have risen to $980 \times 10^6$ therms/day.

Thus all sources, including gas from Alaska, LNG imports, and such gas from coal as there may be, will, it was estimated, be able to supply only

$690 \times 10^6$ therms/day, leaving a shortfall of about $190 \times 10^6$ therms/day.
It was this inability to meet demand, and the possibility of this rising deficit,
which so seriously disturbed the U.S. natural gas industry today.

In view of its dominance as a producer, and especially as a consumer,
of natural gas, it was right to examine the position of the industry in the
U.S.A. However, this examination should be directed not to how the
industry developed in the past, but to what was happening then and would
happen in the future. For many years, gas in the U.S.A. was a by-product
and was sold at by-product prices. Many early contracts were made at as
low as 1 cent/MCF (0.025p/therm) and long-distance transmission lines
brought gas at 5 cents/MCF in the 1930s. Since 1954, the FPC had
regulated these prices and for some time they had re-pegged at a level
which was now about 26 cents/MCF to the producer, which resulted in
about 45 cents/MCF wholesale. Prices for imported gas were about
$1/MCF, which was about the same order as the costs of manufacturing
gas, of natural gas quality, from naphtha, oil, or coal. Efforts to fill the
deficit to which he referred meant that more and more of the gas
consumed in the U.S.A. would have to be provided at this higher level.

This increasingly marked imbalance in gas prices must be viewed against
events elsewhere. Mr Tugendhat referred to the increased price of
80 cents/MCF that the Japanese were to pay for Brunei gas. There were
also negotiations for gas from Iran at a reported 75 cents/MCF, lower
because of the Japanese participation in development. Alaskan LNG was to
rise from 52 cents/MCF to 75−80 cents/MCF in future contracts, possibly
starting that year. All this suggested to the producer in the U.S.A. that a
realistic price for natural gas was not far from $1/MCF. The cry was, "Give us
$1/MCF and we can discover within the U.S.A. all the gas it needs to make
it self-sufficient."

An important development had just occurred. The FPC was considering
a rise in the controlled price from 26 cents/MCF to 35 cents/MCF. If this were
brought into effect, it would mean the thin end of a wedge at the base
of the U.S. gas price structure, which with a few more well-struck blows
would force the price structure upwards towards the $1/MCF goal. Its
effects could not but be felt worldwide.

In conclusion, he had only to draw attention to the competitive
advantage of gas as a clean fuel. Mr Tugendhat referred to it, and there
was to be a complete paper on this important aspect. He mentioned
it because it was certainly a factor in the future climate for exploration
for natural gas.

*Dr J. Birks* (British Petroleum Co. Ltd), in the chair, suggested that from
U.S. exploration discovery experience, if extrapolated on a world basis,
total gas reserves of the order of 10,000 trillion cu ft could be possible.

*Miss M. P. Doyle* (Esso Petroleum Co. Ltd) asked if the Natural Gas Act
of 1954 had affected subsequent extent of discoveries. *A. F. Fox* replied
that he was unable to answer this question.

*J. M. C. Bishop* (Phillips Petroleum Co. Ltd) questioned the advisability of massive investment to market gas in Europe when the demand in the U.S.A. offered a possibly better return. *A. F. Fox* answered that the various possibilities would need to be considered and it was possible that the answer was to export European gas to the U.S.A.

*A. R. Khan* (Gas Development Corpn) questioned the demand estimates given by E. V. Corps for 1980 and pointed out that these assumed a "business as usual" view which he considered unrealistic, a more likely value for this date being 675 with an implied growth rate of about 3 per cent. An exchange between Mr Corps and Mr Khan on sources of statistics (FPC or FRC) left them in disagreement.

*J. F. Allcock* (The Gas Council) said they were indebted to Mr Fox for his paper, which had been read with interest by The Gas Council and would be studied in more detail than they had time to do before the conference. Naturally, one main interest was in the conclusion that the proper price for his hypothetical five trillion field would be 2.3−2.5p.

The Gas Council had a duty, which was not easy to discharge, to settle prices which were highly profitable to the producers in order to generate cash for future exploration and in this way provide a proper incentive for future discoveries. But, as Mr Tugendhat had acknowledged in his keynote speech that morning, they must also have regard to the national interest as well. They had to remember the consumer as well as the producer. They had settled prices in the past which met these objectives and they would in the future.

They were not bound by prices in past contracts but nor were they much moved by cost calculations which attributed enormous exploration expenditures to the discovery of one field. Nor did they concede that when the prices of competing fuels increased, the whole of the margin belonged to the producer.

Incentive to explore was not only a matter of the price. Mr Spalding had spelled it out in his paper. The economic return depended also on the build-up, the depletion rate, and the load factor − and, he added, the tax position of the company concerned. The Gas Council treated each situation on its merits. As his redoubtable boss, the Controller of Purchasing, often put it, "a small field in the Arctic Circle is a different proposition from ten trillion just off Scarborough Pier." For this reason, no two existing contracts in the North Sea were the same and that was likely to remain the case. It was in itself an incentive that the producers knew that whenever they turned up with their drills they would consider the price and terms on the merits of the case.

But in addition to that, Mr Fox had acknowledged that large quantities of associated gas had been found and would yet be found. The producers had incentives to search for oil which were unrelated to the price of gas.

Now Mr Fox sought to establish a supply price for North Sea gas. He told them first that an expenditure of £145 million had discovered 28−30 trillion in the North Sea. But when he came to his cost calculations

he attributed an expenditure greater than the total in the North Sea
to date to the discovery of one five-trillion field. Mr Fox would
appreciate that if he had doubled the exploration expenditure he could
have justified a higher price still. He had drawn attention only to the
most obviously unacceptable assumption underlying his estimate.

But then at the very end of his paper Mr Fox had changed his stance
completely. Having, somewhat surprisingly, approached the price
question by an estimate of cost, he referred to the hardening prices of
competing fuels. It may have been immediately evident that there was a
"margin over the realized price of gas". It may have been evident that as a
result of the Teheran and Geneva agreements, this margin would increase, but
it was not at all evident that the whole of this increase belonged to the offshore
producer. They did not concede this. If it were conceded, then it would
perhaps be appropriate for the Government to consider a royalty rate
which also fluctuated with the movement in the price of alternative
fuels and it might be appropriate for The Gas Council to modify in future
contracts the take or pay obligations it had hitherto accepted.

In reply, *Mr Fox* accepted much of Mr Allcock's criticism, but pointed
out the difficulties of the field producer and suggested that the price of
alternative fuels could not be ignored.

# Supply Aspects of Natural Gas as Affected by Contract Quantity and Quality

## By G. G. SPALDING

*(Amoco (U.K.) Exploration Company)*

---

## SUMMARY

The object of this paper is to discuss the various factors associated with development of a potential gas field. These factors, which are discussed in the paper, are the initial producing rates and build-up period, the plateau producing rate and associated period, the quality of the product, the load factor, and a realistic price which must be attractive to both purchaser and seller. In addition, the paper discusses possible treatment that may have to be carried out on the gas before either transmission ashore or delivery to the market; these include the treatment required to remove water or heavy hydrocarbon.

An example of the procedures that may be followed when developing a small offshore gas field is given. The example, which shows the development schedule and estimates of the capital investment, is for a relatively small offshore gas field with an estimated gas-in-place of 1 trillion cu ft and a total recoverable reserve of 0.7 trillion cu ft, situated 50 miles offshore in 150 ft of water. The development schedule and estimates of the capital costs involved are shown.

*Note:* In this paper a trillion is $10^{12}$.

## INTRODUCTION

The development of a proven gas field depends on many factors, all of which must be considered in order to assess the economic value of a project. The primary object of this paper is to describe the various factors and, secondly, to outline the procedure that may be followed to develop a small offshore gas field.

## FACTORS AFFECTING SUPPLY

The commercial development of a gas reservoir is influenced by a vast number of factors, some of which are partially controllable and some of which are not.

Once a field has been discovered, the sales negotiations hinge on two main points — the price and the quantities to be supplied. Generally, the sooner the producer can recover the reserves — on a fix price contract, at least — the more attractive the economics. However, there are practical limitations: it is not physically possible to deplete a reservoir instantaneously, and it is not possible to dispose of all of the gas immediately. A compromise has to be reached which is realistic for the distributors, in as much as they can expect to sell all the gas which they contract to take, and for the developer, in as much as the actual field development depends to a great extent on the reservoir mechanics. Clearly, if annual contract quantities are to be geared to the volume of reserves in the field, then the bigger the field, the greater the impact it could have on the market. Large fields, therefore, may generally be expected to be depleted over a longer period than small fields. However, if the depletion period is long, the project becomes economically unattractive under a fixed-price system; also, if the pay-back on the project·is extended, the risks are increased.

In some countries, as in the U.K., there is a partial or total monopoly in gas distribution and so the distributor is able to control the demand for gas to some extent. This can be achieved through its marketing activities, by trying to obtain an optimum mix of domestic, commercial, and industrial consumers, and by negotiating long-term contracts with some substantial consumers. A distributor can help to stabilize the pattern of supply by obtaining interruptible contracts with some customers, and introducing peak-shaving storage centres.

The costs and risks associated with a field's development need to be considered, particularly the risk of damage to production facilities and pipe-lines in offshore fields. A large field, to which more then one pipeline is laid, and in which there are several platforms, is more able to bear accidental losses than a much smaller project, for if a pipeline is disrupted, or if a vessel collides with a production well platform complex in a large field, causing irrepairable damage, the replacement cost, although enormous, may be justifiable in relation to the total value of the field, whereas the loss of a platform in a small field, which has been partially depleted, may mean the loss of the entire field. So small fields need to be less exposed to risk than large fields. One way of reducing the risk is to ensure a rapid return on investment; this may be done by increasing the price of the gas and reducing the depletion period, thereby recovering the gas more rapidly.

From a profitability standpoint, initial rates of production should build up to the maximum rate as rapidly as the buyer and seller can realistically manage. Generally, the build-up period will be longer with larger fields due to the problem of absorbing large quantities of gas in the market.

The geographical location of the potential field must be considered; a gas field discovered within close proximity of a market is obviously far more attractive than a field of similar size discovered in a non-industrial area.

Also, due to the lower investment and operating costs, a field discovered onshore would be more attractive than an equivalent offshore field.

## OPTIMUM FIELD PRODUCTION RATES AND CONTRACT REQUIREMENTS

If an operator discovers and delineates a potential gas field, the results of which indicate that the field may be commercial, he will approach potential customers with a view to negotiating an acceptable price and production rate for the gas reserves discovered. (The operator will have already invested considerable sums of money in drilling the exploration and delineation wells and at this time there is no guarantee that a return will be obtained on this investment.) The various factors that have been discussed earlier that affect the development are elaborated on below.

### Production Rates

The demand for gas will vary from a period of high demand in the winter to a period of low demand in the summer. The producer must drill sufficient wells and install suitable producing and treating equipment to deliver the required volume of gas during the peak winter period. A typical sales contract may require the delivery of a specified volume of gas during a year; this volume is referred to as the "Annual Contract Quantity" (ACQ) and the "Daily Contract Quantity" (DCQ) is the ACQ divided by 365 (366 in leap years). During the period of high demand the DCQ will be considerably exceeded, and during the period of low demand the producing rate will be less than the DCQ. Fig. 1 shows how the producing rates may vary during a year. The maximum daily producing rate is

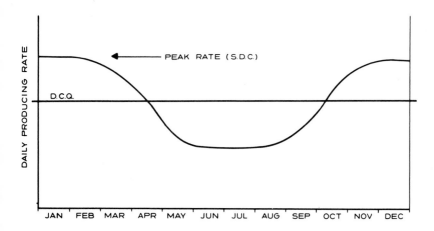

Fig. 1 Typical variation in gas demand

the rate at which the producer must be able to provide gas and in the U.K. is generally termed the "Seller's Delivery Capacity" (SDC), and it is established as a percentage of the DCQ. This percentage will be referred to in this paper as the "load factor". In many countries, load factor is defined as the average daily rate times 100, divided by the maximum daily rate. In this case the load factor is always less than 100.

## Production Schedule

Fig. 2 shows a typical production schedule for the development of a gas

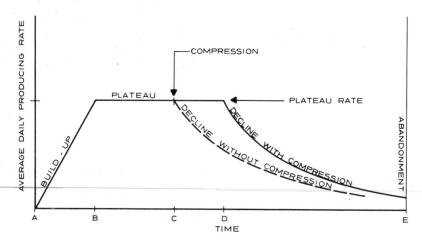

Fig. 2 Typical production schedule

field. The schedule may be split into three distinct time periods as follows:

1. **Period A to B** (Build-up period)

   During this period the developer drills wells and installs the necessary production equipment in the form of pipelines and offshore and onshore treating equipment. The deliverability steadily increases as further wells are drilled and put on stream.

2. **Period B to D** (Plateau period)

   At point B the developer has made sufficient investment to be able to treat all the gas delivered to the purchaser at the SDC rate. During this period the developer would have to drill additional wells in order to maintain deliverability at the maximum SDC rate until a point C is reached where the drilling of further wells becomes uneconomic due to the excessive number of wells required to maintain deliverability. At point C, the developer may install compression facilities in order to maintain deliverability through to point D.

3. **Period D to E** (Decline period)
   A point in time will be reached where the investment of additional
   capital becomes uneconomic (C or D), after which point the deliverability will be limited and will steadily decrease as the reservoir pressure
   declines until the field is abandoned at point E, due to the operating
   costs exceeding the income. (If the installation of compressors at point
   C could not be economically justified, the decline would commence at
   point C).

The developer's investment schedule is directly dependent on the plateau
producing rate, the build-up period, and the period during which the plateau
rate is maintained. Before deciding to commit capital for the development of
the field, the developer must know the plateau producing rate and the
respective periods of build-up and plateau. In turn, the purchaser may also have
to invest monies in order to deliver the gas to his market. Therefore, after the
existence of reserves is proven and prior to commitment, an agreement has to
be made between the developer and the purchaser covering the following
specific factors:
   The build-up period and associated producing rates
   The plateau period and associated rate
   The load factor
   The quality of the product
   The price and price adjustment provisions

**Build-up Period and Associated Rates**

The period of time during which the producer invests the major portion of
capital and the purchaser creates a market, installs a distribution network, and,
if necessary, converts appliances for use with natural gas, is generally related
to the size of a field; production from a small field could be readily absorbed
in an existing system, whereas a relatively large field would have a tremendous
impact on the market. Theoretically, the shorter the build-up period, the
higher the worth of the project; however, in practical terms the period may
vary from two years for a small field to five years for a large field.

During the build-up period, the deliverability steadily increases, the
producer drills more wells, and the purchaser expands the distribution network.

**Plateau Period and Rate**

The most important item that affects the development schedule is the
plateau producing rate, which in turn is dependent upon the reserves of gas
in the reservoir. Obviously, if an extremely high plateau producing rate is
negotiated for a small field, the period for which this rate can be maintained
will be short and the field will be exhausted in a short period of time. This
situation would generally be unacceptable to the purchaser. In some cases the
plateau producing rate may be related to the reserves and an associated period
of time, which is known as the depletion period.

As an example, the plateau producing rate could be determined from the equation:

$$\text{Daily plateau rate} = \frac{\text{original reserves of gas}}{365 \times \text{depletion period (years)}}$$

With normal field production the depletion period is a misnomer, the producing life of the field is generally greater than the depletion period, as production is not maintained at the plateau rate during the entire producing life.

Obviously, the selection of a suitable value for the depletion period will affect the development programme and subsequently the financial attractiveness of the project. The evaluation of the worth of a project will vary between different development companies due to each company having different policies regarding, for example, the design and operation of the production equipment, or taxation. However, by using typical North Sea figures for the development of an offshore gas field (and assuming a fixed price over the yearly producing

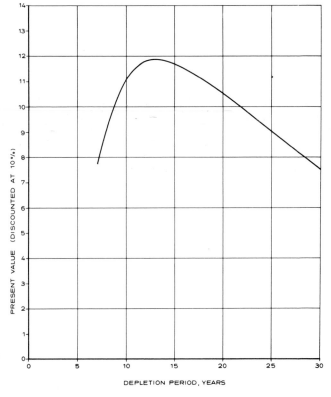

Fig. 3 Effect on project worth of varying depletion period

life), the curve shown in Fig. 3 can be developed. The curve shows the present value of developing and producing a gas field with different values for the depletion period. The most attractive situation from the present value standpoint in this example occurs when the depletion period lies between 10 and 15 years.

The period of time for which the plateau rate is maintained is particularly important to the purchaser. If a specified steady supply of gas is available, the purchaser is then able to negotiate contracts with third parties which will be compatible with the available supply.

Generally the period of time for which the plateau rate is maintained is related to a certain percentage of the reserves. From a practical standpoint, it would be impossible to maintain the plateau rate until 100 per cent of reserves are produced. Fig. 4 shows the financial effect of varying the proportion of reserves produced at the plateau rate; the optimum value lies between 50 and 60 per cent.

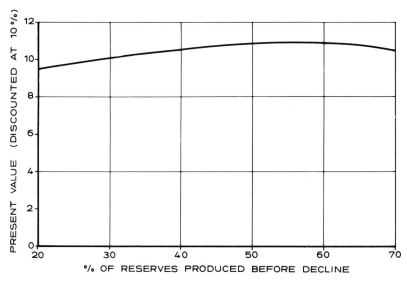

Fig. 4 Effect on project worth of varying percentage
of reserves before decline begins

**Load Factor**

From the producer's point of view, it would be more efficient and would require less capital investment if the field could be produced at a steady rate. Therefore, the producer aims to negotiate a load factor as close to 100 per cent of the Daily Contract Quantity as possible. However, by the very nature of the product, demand for gas fluctuates considerably. The distributor would ideally like to have complete freedom of access to supplies. So, once again, a compromise is reached between the two parties to the contract. The greater the peak load factor, the greater the producer's investment must be, and therefore the less economic the project becomes. The amount of extra capital investment required is directly proportional to the load factor. By applying a load factor, the present worth of the project is proportionally reduced; therefore, in order to provide an incentive to install the additional facilities, the producer will require an increase in the gas price to cover the additional capital expenditure.

## Product Quality

Generally, the gas produced requires some form of treatment before delivery to the purchaser. It may be necessary to provide treating equipment to remove detrimental components of the gas such as:

1. Water
2. Excess liquid hydrocarbons
3. Carbon dioxide
4. Hydrogen sulphide
5. Nitrogen

Water and liquid hydrocarbons are normally present to some degree in natural gas, the remaining three items may not be present in sufficient quantities to warrant treating. In order for the producer to efficiently design the production equipment, the producer and the purchaser must agree on the specification of the sales gas before the final design work is initiated.

## Offshore Processing

The production facilities installed offshore would be designed to process the gas to a specification suitable for transmission ashore. The gas in the reservoir would most likely be saturated with water vapour and hydrocarbons at the reservoir temperature and pressure; if this gas was produced directly into the transmission line, without removing the free water and a certain amount of the water vapour, hydrates would form (a hydrate is a particle of solid matter, usually consisting of methane and water, that is formed as a result of a reduction in the temperature of the gas when in contact with free water). The hydrates would steadily build up until a permanent block in the pipeline formed. To prevent the formation of hydrates, the free water is removed by primary separation, the water vapour may be removed from the gas by contacting it with a desiccant, or methonal may be injected to suppress the hydrate formation temperature. If an absorption process is used, the liquid hydrocarbons are also separated from the gas stream prior to drying, and then injected into the pipeline downstream of the process facilities and carried ashore with the dry gas. In addition to processing the gas for the removal of water, it may also be necessary to treat offshore for the removal or inhibition of carbon dioxide and hydrogen sulphide. The presence of either of these components in the gas in sufficient quantities can be the cause of very rapid corrosion. If excessive carbon dioxide or hydrogen sulphide are present, it may be necessary to either remove the carbon dioxide or hydrogen sulphide, or inject some form of corrosion inhibition to protect the production complex.

## Onshore Processing

To keep operating costs as low as possible, the developer will normally only install sufficient process facilities offshore to treat the gas for transportation onshore. However, before the product is delivered to the purchaser, it may be necessary to further treat the gas to:

1. Remove liquid and some heavy hydrocarbons in order to prevent retrograde condensation
2. Remove excess hydrogen sulphide to eliminate its toxic effect.
3. Remove excess nitrogen, since the presence of nitrogen, in appreciable quantities, will considerably lower the heating value of a gas. However, it should be noted that, in some markets, it may be necessary to inject nitrogen in order to reduce the heating value of the gas to produce a product compatible with existing supplies.

The excess hydrocarbons can be removed by low temperature separation and the excess carbon dioxide or hydrogen sulphide may be removed by a physical or chemical treatment. The removal of nitrogen can only be accomplished by liquefying the hydrocarbons and using low-temperature separating techniques. The developer will design and install sufficient processing equipment to carry out the necessary treatment of the gas to meet the specification negotiated in the contract. The trunk line from the field to the shore terminal would be designed for minimum pressure loss, consistent with reasonable capital investment.

## Price

The price paid for the gas will depend upon all the factors previously mentioned;.before the producer is prepared to invest capital, however, the purchaser and seller must agree on a realistic price which is satisfactory to both parties. In an entirely free economy the price paid for the gas should be competitive with alternative fuels. It should be noted that natural gas has many advantages over the alternative fuels, including pollution effects which, in the case of natural gas, are negligible.

## GAS FIELD DEVELOPMENT

In order to demonstrate the techniques used to develop a gas field, an example is given below. This example assumes that the operator has discovered and delineated an offshore gas field, as shown in Fig. 5, with the following properties:

1. Total gas in place                 1 trillion cu ft ($10^{12}$)
2. Reservoir pressure                 3500 psia
3. Reservoir depth                    10,000 ft
4. Water depth                        150 ft
5. Distance to proposed shore
   terminal                           50 miles
6. Recoverable reserves              0.7 trillion cu ft ($0.7 \times 10^{12}$)
7. Average well deliverability at
   the original reservoir pressure   20 MMCFD
8. Type of production                 Gas expansion

While in most circumstances a field of this size, discovered on land, would be financially attractive due to its large reserves, for an offshore operation an operator would critically analyse the proposed development programme and capital investment, particularly in an environment as hostile as the North Sea, before committing funds to the project or negotiating an acceptable price.

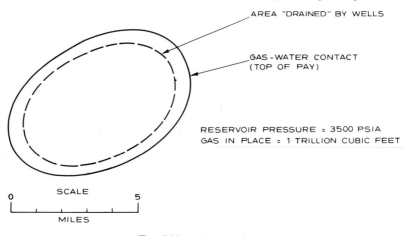

Fig. 5 Hypothetical field

Based on the factors discussed earlier, the operator could design a development programme that is based upon a depletion period of 15 years and a load factor of 150 per cent. The plateau production rate will be maintained until 50 per cent of the reserves have been produced.

The calculated plateau rate will be

$$= \frac{0.7 \times 10^{12}}{365 \times 15} = 128 \text{ MMCFD}$$

and the maximum deliverability will be
= load factor x plateau rate/100
= 192 MMCFD

Ignoring compressibility factors, the reservoir pressure at the beginning of the decline period will be 2275 psia. Table I below shows the well deliverability at the beginning of the decline period and the total number of wells that would be needed to provide the required deliverability with and without compression.

TABLE I

| Case | Estimated well Deliverability (MMCFD) | Wells Required |
|------|---------------------------------------|----------------|
| No compression | 6.2 | 31 |
| Compression | 10.1 | 19 |

In all probability, at least one stage of compression would be installed, therefore the maximum number of wells that would be required is 19. The total area to be drained by the wells could probably be less than the actual area enclosed within the gas water contact. Fig. 6 shows how the area drained

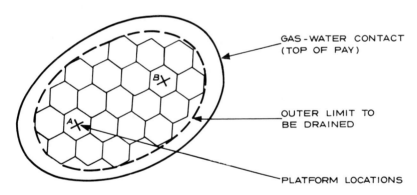

Fig. 6 Hypothetical field drainage pattern

can be envisioned as being split up into 21 hexagonal areas, each of which could be drained by a well. (The effective radius of drainage of each of these areas is 3000 ft). There are two alternative methods of drilling wells to the bottom hole locations. Either (a) erect permanent platforms and drill several directional holes from each of these platforms, or (b) drill vertical holes at each location (with the wellhead installation on the sea bed) and connect each well to a central gathering station. In the example the water depth is shallow enough to allow either type of development. The industry has developed proven techniques for the development of a field from fixed platforms; however, the problems associated with the completion of wells at the sea bed in hostile environments are considerable and generally economically less attractive than wells drilled from permanent platforms; the field will, therefore, be developed from fixed platforms.

Fig. 7 shows that the maximum "step-out" that could be drilled from a fixed platform would be 10,000 ft, assuming an average deviation of 45°.

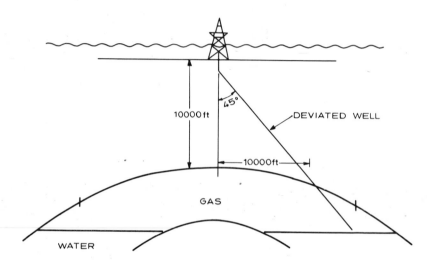

Fig. 7 Hypothetical field, maximum distance that
can be reached from a fixed platform

Using this maximum value of "step-out" distance, it would be possible to drill to ten bottom hole locations from each of two platforms situated at points A and B, as shown in Fig. 6. The design of the well completion equipment and casing/tubing sizes would depend upon the volume of gas to be produced. A typical casing programme is shown in Fig. 8. In the event of a disaster, such as a vessel colliding with a platform or a fire on a platform, provision is made to install a downhole surface-controlled safety valve in each well. The valve would be of a fail-safe design and would, in the event of a disaster, shut off the flow of gas from each well.

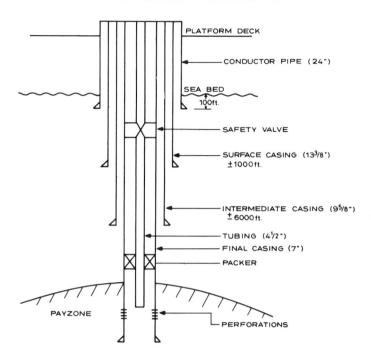

Fig. 8 Hypothetical field, example of casing strings

Fig. 9 shows the theoretical well schedule. On average, during the early part of the production period, two or three additional wells would be required each year. The cost of mobilizing and demobilizing a drilling rig and associated lift equipment for drilling two or three wells from a platform per year would be prohibitive. The drilling schedule would, therefore, be accelerated and nine or ten wells would be drilled consecutively from one platform; drilling from the second platform would then be scheduled so as to provide sufficient extra deliverability when sufficient gas could not be produced from the first platform. The investment and development schedule is shown in Fig. 10; the investment costs shown are based on typical North Sea values. The operator can, with reasonable accuracy, estimate the capital investment costs. However, in addition to the capital investment, the operator will also have to consider operating costs. Provisional estimates of operating costs can be made on the basis of costs incurred in producing other similar fields. Operating costs are normally split into (*a*) direct operating costs, which include maintenance costs, consumable costs, and personnel costs

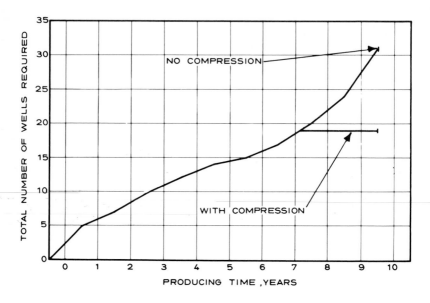

Fig. 9 Hypothetical field, well requirements (GIO = 1 trillion cubic feet)

| ITEM | YEAR OF PRODUCTION | | | | | | | | | | | | |
|---|---|---|---|---|---|---|---|---|---|---|---|---|
| | -2 | -1 | 0 | 1 | 2 | 3 | 4 | 5 | 6 | 7 | 8 | 9 | 10 |
| EXPLORATORY DRILLING | ⊠ 2000 | | | | | | | | | | | | |
| DELINEATION DRILLING | | ⊠ 1200 | | | | | | | | | | | |
| DRILLING PLATFORM | | | ⊠ 1000 | | ⊠ 1000 | | | | | | | | |
| DRILLING WELLS | | | 3 1200 | 7 2800 | 7 2800 | 2 800 | | | | | | | |
| TRUNK LINE | | | ⊠ 7200 | | | | | | | | | | |
| PROCESS PLATFORM | | | ⊠ 1000 | | ⊠ 1000 | | | | | | | | |
| SHORE FACILITIES | | | ⊠ 1000 | | | | | | | | | | |
| LINK LINE | | | | | ⊠ 400 | | | | | | | | |
| COMPRESSION | | | | | | | | | | ⊠ 1000 | | ⊠ 1000 | |
| TOTAL ANNUAL INVESTMENT | 2000 | 1200 | 11400 | 3800 | 4200 | 800 | 0 | 0 | 0 | 1000 | 0 | 1000 | 0 |

COST SHOWN ARE £000

Fig. 10 Hypothetical field, investment and development schedule

associated with the production of gas, and (b) indirect operating costs, which include overheads such as office rental, administrative costs, and indirect service cost. For the example shown, a typical operating cost may be of the order of £1 million/year for the life of the field. By analysing the investment and operating cost, the operator is able to estimate the economics of a given gas sales proposal. Obviously, different operators will have different criteria for investment, giving due recognition to the risk involved. The operator may have invested considerable sums of money in exploring without success other prospects within the area. The cost of the exploratory wells, together with the cost of the delineation wells, will have to be recovered and would normally be offset against the project.

It is highly unlikely that the performance of the reservoir will follow a straightforward formula; during the early producing life of the field the operator will, by performing certain tests, be able to predict more accurately the future behaviour of the reservoir. Changes in the performance may require the drilling of additional wells or the acceleration of the compression facilities.

## CONCLUSION

In conclusion, it can be seen that several interrelated factors must be considered before the decision to commit capital to develop a gas field is made. In areas of high risk, such as the North Sea, an operator will be reluctant to invest capital in a project unless it yields a return equal to other investment alternatives and substantially in excess of that available through low risk investments at the prevailing interest rates.

## ACKNOWLEDGEMENT

The author wishes to thank the management of Amoco (U.K.) Exploration Company for permission to publish the paper. It should be noted, however, that the author's opinions and comments contained within the paper do not necessarily reflect those of Amoco.

## DISCUSSION

The discussion on this paper was opened by G. T. S. Cribb (The Gas Council), who raised the point that a rapid build-up of offtake rate was possible for North Sea gas because of the facilities previously installed to distribute imported Algerian natural gas. With regard to load factors, superimposed on the seasonal smoothed rates were daily fluctuations, leading to needle peaks of demand, but 60 per cent was contractual load factor for minimum seasonal normal temperature. Peak demand load factors could go below 50 per cent, but were held to 60 per cent by peak shaving with LNG. The North Sea fields were fortunate in having a gas free from hydrogen sulphide, save for one formation in the Hewitt field, and the gas quality specifications had resulted in a gas with a Wobbe number close to that of pure methane. Whilst this specification may have appeared unduly onerous, it was in the interests of both producer and distributor that the transmission and distribution system was trouble-free in operation. In conclusion, although natural gas had undoubtedly many plus features over alternative fuels, when it came to storage as LNG this was not the case, and considerable opposition had to be faced on environmental/safety grounds at proposed LNG sites.

*J. Prescott* (British Petroleum Co. Ltd) enquired if the load factor could not be improved by the more extensive use of LNG facilities. Mr Cribb replied that if LNG facilities were not being used, the load factor would be far worse. *J. F. Allcock* (The Gas Council) was still not clear on load factor *vis-à-vis* plateau rates and percentage of reserves recovered. He understood the paper to show that the plateau rate could be extended until 50 per cent of reserves were recovered. *Mr Spalding* replied that, contractually, 40 per cent of reserves were to be produced at plateau rates, but this could be extended to 50 per cent if the economics of installing additional compression were made attractive.

*J. M: C. Bishop* (Phillips Petroleum Co. Ltd) found the direct relationship between load factor and price hard to follow. *Mr Spalding* replied that if a higher than normal load factor was required for a limited seasonal period, requiring extensive re-vamping of facilities, then an increase in price was necessary to cover the cost of these facilities, which would be under-utilized most of the year.

Mr Bishop also suggested that intensive depletion of a field in five years should be more attractive than 15 to 20-year depletion. Mr Spalding replied that the enormous capital required for a high-volume, short-life scheme made the present worth of such a project unattractive.

# Economics of Long-Distance High-Pressure Gas Transmission

By N. W. ROBERTS and M. J. STEWART

*(Constructors John Brown Ltd)*

## SUMMARY

A fundamental difference between the economics of pipelining gases and liquids arises from the fact that gas is compressible. While with liquids the pressure drop is constant along the line between pumping stations, with gases the pressure drop per unit length rises as the pressure falls. This suggests that, in order to reduce the cost of gas transmission, pressures should be as high as possible; however, material availability and welding technique limit the pressures which can be used in practice. By an analysis of pressure losses, the incentive for improvement is defined. The ultimate increase in density is achieved by piping LNG, and the characteristics of such a scheme are discussed.

Few pipelines operate under constant load conditions throughout their lives and the economics of any single pipeline scheme will depend upon the accommodation for future demand which has been built in initially. The various factors by which an ever-growing market can be accommodated, together with the economic time intervals for new construction, are discussed. These time factors are likely to be influenced by environmental conditions.

One of the more powerful means for improving the economics of pipeline transmission is the provision of storage in gaseous or liquid form. The various means of providing storage are discussed.

## INTRODUCTION

At the outset it is desirable to define the objectives of this paper in relation to the other papers in the symposium. First, it is a speculative paper; it deals with a number of ways in which current practice might develop. Secondly, it is a paper about economics and in no sense a do-it-yourself pipe-lining manual. Thirdly, it is written as part of this symposium, which is convened for the purpose of discussing alternative methods of transporting natural gas.

The chief method, competitive with pipeline transmission, for transporting natural gas is as liquefied natural gas in tankers. The application of either method is subject to certain characteristic constraints which need to be taken into account as the first stage of any assessment of relative merits. The nature of the constraints is such that one or other method may well be ruled out. The constraints may be summarized in Table I, which in either case refers to a new scheme.

## TABLE I

| Long-distance pipeline | Liquefield natural gas |
|---|---|
| Heavy initial capital cost. | Heavy initial capital cost. |
| No initial conversion cost or compression cost. | Heavy cost of liquefaction. |
| Minimum purification cost. | Stringent purification requirements. |
| Moderate re-compression costs. | Low unit transportation cost. |
| No lower limit on throughput. | Defined minimum throughput in relation to unit output. |
| Ceiling limit of throughput | Upper limit by unit output. |
| Modest extensibility through additional re-compression stations at modest cost. | No extensibility without complete liquefaction unit or liquefaction tanker combination. |
| Slight inherent storage facility. | Unlimited relatively cheap storage at either end. |
| Inflexible source of supply and delivery. | Ability for both buyer and seller to switch source of supply or delivery point. |
| Overland route with short sea or river crossings. | Both supply and delivery point with access to sea route of competitive length. |

Within the boundaries of these constraints, we may compare the two possibilities for any given situation. The comparison, however, is subject to one very important geographical constraint, which is that the sea route and land distances are almost never equal. For this reason, every comparison is individual to its case.

For short distances between source and demand, the pipeline is the only choice. Indeed, if we assume a cost of liquefaction of natural gas of 10p/ million Btu of gas then liquefaction will not be considered unless the pipeline transmission cost by the optimum route exceeds this figure. Approximate data based on present-day practice suggest a consequential radius of 1500-2000 miles. The very low comparative cost per mile of tanker transport suggests that this radius will be relatively unaffected by the distance ratio between the routes.

At this point a note of caution needs to be sounded with regard to overall quoted pipeline costs. The *Oil and Gas Journal* has published[1] a comparison of oil pipeline construction costs which suggests that there is a standard deviation of ± 40 per cent between the costs of otherwise similar pipelines. Costs which are assumed or realized by particular Engineering Executive Groups* will vary more than these because the final costs are influenced by the selective tendering process which tends to reduce their variance.

At the same time, the costs of pipe laying are considerably influenced by environmental conditions. These questions, though basic to the economics of a particular scheme, are not discussed here.


## THE INFLUENCE OF DESIGN PARAMETERS

In comparing the economies of gas and oil pipelines, there are some factors which are common to the two problems and some which are peculiar to gas.

For both oil and gas it is a fact that unit transmission costs increase rapidly as the quantity decreases. For oil, the availability of road, rail, or river transport competition sets a constraint, so that oil pipelines are rarely built for small quantities. With gas, however, these alternatives are not feasible so that we do find gas pipelines in all sizes running down to the domestic service. The point is important because the range of throughputs in gas piping tends to obscure the need for different treatment. Standards have been evaluated for underground gas pipes for all duties and fall into several different classes. In a sense these represent different technologies and the long-distance pipeline does not necessarily have to use the same techniques as medium- and short-distance schemes.

The flow through an oil pipeline is considerably affected by the viscosity of the oil, and although this is an important element in the operating cost to a trunk pipeline, it has a once and for all effect on the design process. At the same time, however, it has to be recognized that things change and that during the life of a pipeline it is not unknown for crude characteristics to change. With gas, however, this effect is negligible since natural gas is much more uniform in quality. A strong factor in creating this situation is that it is usual to purify natural gas supplies at source to full user standards, removing useless load and also making possible offtakes for subsidiary users along the length of the pipeline. At the same time, it must be recognized that natural gas purified to the standards required for gas transmission falls short of the standard of purity required for liquefaction.

Hubbard[2] has studied in great detail the economics of oil pipeline operation, primarily for comparison with other forms of oil transport. In particular, he has

---

* An Engineering Executive Group is a group of engineers charged with the realization of a pipe scheme, whether employed by a contractor or by his client.

evaluated average costs of operation as they vary with load and design parameters, *e.g.* selected diameter, pump station spacing, and so forth.

In the course of his discussion he presented one conclusion which is so important that we reproduce from his paper Fig.1, which demonstrates convincingly that the lowest cost of transporting oil by pipeline does not lie at the load points corresponding to the minima for individual diameter curves. In other words, when operating a pipeline of a given diameter, one can get a lower unit transportation cost by forcing more oil through it, but when the pipeline exists only on paper, even lower costs can be achieved by building in the next largest available diameter.

### Fig. 1

Transmission cost v flow rate

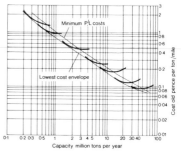

When we come to explore whether similar concepts apply to gas pipelines, the problem acquires an additional dimension. In an oil pipeline the friction loss at a given rate of flow is virtually independent of the pressure. Thus the operating cost of the line between pumping stations is a linear function of length (subject to hydrostatic correction terms, which are a function of route countours). The placing of pumping stations, both for initial operation and for subsequent capacity expansions, is a simple decision choice. The thickness of the pipe which is laid will often be decided by the need to resist hydrostatic pressure when flow is interrupted, or by surge considerations, rather than by pumping head.

However, in a gas pipeline at constant diameter and load, the friction loss varies to a first approximation inversely with the square of the pressure. Also friction loss results in pressure reduction, so that the operating cost of a gas pipeline can only be obtained by integration of the friction loss from point to point. This has two effects on the economics. First, there is a fairly definite lower limit to the pressure which can be tolerated if flow is to be maintained; and secondly, costs decrease as the operating pressure is raised.

Chiefly, this arises because the density of natural gas even at comparatively high pressure is so much lower than that of oil and decreases further as head is lost. Velocity, too, at constant mass flow increases as head is lost. The combination of these two factors sets a definite lower limit to the pressure which can be tolerated and dictates a minimum spacing of the pumping stations if costs are to be kept within balance. A more favourable factor is that the lower density of gas means that the hydrostatic effect of topography is considerably less and is rarely a factor with which to be reckoned.

The nature of the relationship for restoration of lost head is also different for gas and liquid. In an oil pipeline, pumping power is linear with lost head. For gases the relationship is more complex but for understanding we may quote the ideal gas equation.

$$HP = C.RT \ln (p^1/p^2)$$

Where $HP$ is the horsepower requirement
$C$ is a constant related to flow rate
$RT$ is the gas constant multiplied by the absolute temperature
$p^1/p^2$ is the pressure ratio

In terms of operating cost, therefore, head lost at higher pressures is substantially cheaper than head lost at lower pressures. However, in evaluating the statement it should be borne in mind that few users require natural gas to be at pipeline pressures, even at today's levels, and that power expended to produce unnecessarily high terminal pressures is wasted power.

Nevertheless, there is a case for investigation whether the use of higher pressures in long-distance pipelines, where terminal effects are in any case small, can lower the cost of long-distance transmission of natural gas.

## PRESSURE DROP INVESTIGATION

It seemed desirable to us, for the reasons which have been outlined, to investigate whether there would be any economic advantage in increasing the working pressure of long-distance pipelines. We appreciate that such a proposal would call for considerable changes in current practice; at the same time, such changes would be unlikely to occur unless there were an incentive to make them. This section presents the results of our investigation and they are discussed in the following section. Details of the calculations which we made are given in the Appendix.

We have already discussed and illustrated the inherent variance of pipeline costs. Thus, although the costs presented here bear a true relationship to one another, all being based upon the same basic assumptions, they cannot be considered as an actual cost of actual pipelines. In other words, the graphs cannot be used as an instant pipeline costing device! The basic costs used, which are described in the Appendix, are general figures which have been formulated from true cost figures arising from CJB's experience.

We have taken this course because we believe that the trends are better demonstrated in this way. Secondly, as described earlier, there are so many parameters affecting the actual costs of laying gas pipelines that it would be an impossible task to consider and analyse them all. The results of an analysis would have little more than local significance. Consequently, to formulate the cost here, several simplifying assumptions have been made and all basic cost data used are averaged.

It is worth noting here the major assumptions which have been made:
1. The gas flows at a constant temperature – this, although nearing the truth for large diameter lines, is only partly true in small diameter lines (but only after the gas assumes ground temperature).
2. The ground is flat.
3. The cost of purchasing of "rights of way" has been neglected, as this is one factor which would not lend itself to generalization.
4. The operating costs have been constructed by computing the pressure drops down an increment of length of a pipeline and, by so doing, the specific problems of positioning of compressing stations have been overcome. The cost of restoring lost pressure is thus an average figure which has been based on average compression costs and does not assume any particular size of machine.
5. The pipeline costs used were actual costs for the diameters and thicknesses which, however, are step functions. Similarly, the laying cost of the pipe is a step function which has been based on figures formulated from CJB's experience.
6. One very important point which greatly affects the economics of a pipeline is the "write-off" period of a pipeline installation – this has been taken as ten years, although shorter periods may be appropriate when natural gas is penetrating a new market, as discussed below.
7. The other costs which complete the total cost of gas transmission per quantity and distance are again computed by the use of smoothed but varying percentages.
8. No allowance has been made for return on capital other than write-off.

We appreciate that such a basis would not be appropriate for the design of a specific pipeline. Since our intention is to stimulate thought and discussion, and since we are not aware of previous work on this topic, we felt that the approach was justified.

## DISCUSSION OF RESULTS

All the results presented in this paper are given in graphical form, as they are intended to demonstrate trends and generalities only.

Fig.2 is included mainly to enable a direct comparison to be made with that of Hubbard [2] for oil pipelines (a copy of Hubbard's graph is given as Fig.1). It can clearly be seen that the form is basically the same, although, in the case of gases, it has been necessary to eliminate the parameter of pressure by considering a fixed inlet pressure. The main reason for making the comparison is to show that, as for oil pipelines, the optimum operating point for any diameter with respect to flow is not necessarily the most economic point for transporting that flow if it is feasible to consider other diameters. The envelope curve can be used to locate the most economic operating system, for the fixed inlet pressure, for any flow.

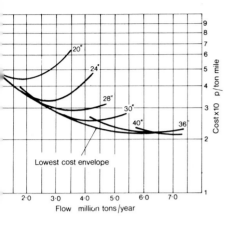

Fig. 2

Transmission cost v flow rate (at a constant inlet pressure of 1000-1050 psia)

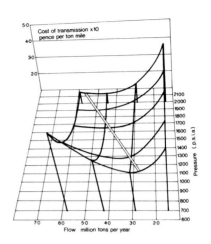

Fig. 4

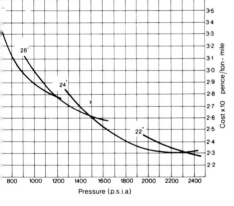

Fig. 3

Transmission cost v inlet pressure (at a constant flow of 2.86 million tons/year)

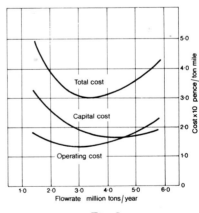

Fig. 5

Transmission cost v flow rate (showing the breakdown of total cost for a 28-inch line at about 900 psia)

A second two-dimensional plot is given in Fig.3, which shows how the cost of transporting gas varies with the line pressure, at constant flow. Again, we see that the most economic operating points do not coincide with minimum points for individual diameters if, in fact, any minima are ever reached.

Each of Figs.2 and 3 shows only a very limited picture but a series of curves for a 28-inch diameter line have been combined to give Fig.4. Fig.4, which is a diagram of a three-dimensional model, attempts to display the surface of cost as it varies with flow and inlet pressure. The lines on the surface which run from the top to the bottom of the figure are lines of constant flow and the lines which cross the surface from left to right are at constant pressure. This is the simplest picture we can present of the economic relationships and shows how the transmission costs fall, as expected, with increasing pressure. The broad white line locates the minimum cost for any flow and shows how this minimum reduces, and occurs at increasingly high flows, with increasing pressures.

Figs.2, 3, and 4 show cost trends without any indication how the cost is made up of operating and capital costs. Fig.5 gives this breakdown for one section through the surface of Fig.4. The same could be plotted for any section of Fig.4. It does seem from Fig.4 that, in the range quoted, increasing pressures will continue to reduce overall costs. The practicalities of such pressures are going to be the immediate limiting factors, though the compressibility will impose an ultimate limitation. For these costs to decrease, however, it does mean that the cost of the high-pressure pipe does not increase out of all proportion to the present cost to diameter ratio. For example, by extrapolation from existing costs the cost of transporting gas in a 40-inch pipe at 2000 psig would be 0.14p/ton-mile, saving about 0.08p/ton-mile over the cost for conventional 1000 psig 40-inch pipe.

Applying a similar analysis to diameter, a forecast saving of 0.03p/ton-mile could be achieved over 40-inch pipe by increasing diameters to 60 inches.

In a sense, the concepts developed here are in current practice whenever a pipeline is laid which is operated at a throughput well below its ultimate load. The operating cost can be kept to a minimum by operating at the full eventual working pressure. If, however, the pipeline were designed for a higher ultimate pressure, with consequential smaller diameter, even more advantage could be gained and this would persist throughout the life of the pipeline.

## THE CONSEQUENCES OF INCREASING OPERATING PRESSURE IN LONG-DISTANCE NATURAL GAS PIPELINES

If there were to be a move towards substantial increases in the operating pressure of natural gas pipelines, a number of factors would have to be looked at in detail before it could be agreed that such a change would be practical.

The first factor of an enquiry relates to the availability of the necessary high-pressure rotating machinery for the compression stations which would be required. Ten years ago this would have been a complete barrier to any such suggestion, but since 1965 rotary compressors have been available compressing synthesis gas from 25 to 150 atmospheres. As with any radical new engineering departure, some of these machines had their teething troubles, but over 100 are now in satisfactory operation throughout the world and the time is surely not far distant when their reliability will be deemed sufficient for pipeline transmission service.

The past 15 years has seen continuing increase in the availability of higher tensile steels for line pipe. On occasion, some of these specifications have presented difficult field welding and inspection problems, but there is no reason to doubt the ability of the industry to cope with further improvements as they become available.

Nevertheless, even given conceivable increases in steel quality beyond the X100 steel currently under evaluation, some increase in thickness would be necessary. From the construction point of view this is unlikely to pose serious problems. Handling would be in some respects easier, since the additional stiffness might make a pipe more forgiving. Welding costs would increase, but probably not in proportion to the weight of pipe. The use of automatic welding machines would be encouraged. The advent of crawler x-ray machines has meant that inspection requirements can be met.

The manufacture of the pipe which would be required would present considerable difficulties which are unlikely to be resolved until the requirement is clear.

The safety problems raised by an increase in operation pressure do not differ in kind but only in degree from those encountered at present.

The final problem is one of physics. Natural gas is primarily methane, and although its liquefaction point is at a very low temperature, it is still at a substantially higher temperature than the so-called permanent gases. At very high pressures, of the order of 100 atmospheres, its properties begin to partake in some slight degree to those of a liquid. The applicability of conventional flow formula is thus called into question and it can well be suggested that the industry should begin to examine conventional practice in this respect.

One logical conclusion of this argument is to transmit the gas at the maximum density which is possible, i.e. as liquid. This possibility has been discussed, most recently by Katz and Hashemi[3]. They have studied in great detail the physical and economic factors involved and conclude that the proposal warrants further study. However, their analysis appears to apply only to the case of overland transport of available LNG, i.e. a straight comparison of pumping LNG as against gasification and pipeline transport. The economic advantages which they claim do not appear to be sufficient to pay for the cost of liquefying natural gas for the purpose of pipeline transport.

We may remark in passing that a modest cooling of pipeline gas could result in a worthwhile increase in pipeline capacity, although it would take many months to cool the ground surrounding the pipe to the required degree.

## LOAD FACTORS

Two forms of load factor need to be considered in arriving at an economic design of a pipeline. The first of these factors is the provision which needs to be made for the growth of the market which the pipeline needs to serve during its life. The second is the provision for seasonal and daily variation in the load which the market presents.

When considering a new pipeline installation, the nature of the market it is designed to serve has to be thoroughly analysed. This problem is dealt with in other papers and the purpose of this brief discussion is to highlight the factors directly affecting design. If the market is a developed industrial region to which natural gas is being introduced for the first time, experience throughout the world suggests an extremely rapid growth which derives from the marked preference for natural gas by fuel users; the growth rate invariably seems to exceed the provisions of the planners. Odell[4] has discussed the growth demand from the Groningen field, which appears to have a growth rate of 20 per cent/year. This growth rate applies to the Dutch national market; the export market appears to be growing at an even greater rate. He suggests that by 1975 at least 52 per cent of the Netherlands total energy consumption will be provided by natural gas. Most of this growth has been due to substitution for other energy sources and when the substitution is complete the growth in demand should reduce to the normal 3 per cent growth for the country's energy demand.

It follows that when a transmission pipeline is being built for introducing gas into a new market, generous provision needs to be made for increases in flow. The cost characteristics of transmission pipelines are such that the provision of such reserve capacity is relatively cheap, particularly when the investment is calculated on a discounted cash flow basis.

However, where substitution is complete, the case for reserve capacity is less strong and we might offer the speculation that the import of LNG in such a case can be more readily justified if LNG in the future will become a universally available commodity in world trade. The discussion by Salkeld[5], for example, views the current massive import programme of LNG into the United States natural gas grid as a decision forced on the U.S.A. by the impracticability of developing further domestic reserves in the time before which they are needed but the decision may well be soundly based in economic

The other factor affecting the load factor of a pipeline is the question of daily and seasonal fluctuations in demand. With an LNG installation, storage is cheap and these fluctuations can be readily made; indeed, Clar *et al*[6] have pointed out the important role which LNG supplies in meeting these fluctuations in the U.K. natural gas system design.

With a long-distance transmission pipeline, daily fluctuations can in part be met by drawing on the reserve of gas held in the pipeline. This practice has its limitations and, in general, if there is need to resort to it new capacity is needed; nevertheless, the facility is there.

For seasonal fluctuations, however, the amount of gas retained in the pipeline is totally inadequate to meet practical load conditions and reserve capacity needs to be built in. Peak shaving facilities have received considerable attention and their economic equivalent of interruptible supplies is already in wide operation. The principal peak shaving facilities include LNG plants at the consumption end of the pipeline, which increase the pipeline load during periods of low market demand and reduce it by re-evaporation of LNG during the periods of peak demand. It has to be recognized, however, that whenever there is a choice LNG manufactured after the gas has borne the cost of pipeline transmission is inherently more expensive than LNG imported from abroad.

Where geographical conditions permit, the use of underground caverns for storage of gas under high pressure can be a useful peak shaving device. However, such conditions are rare.

## CONCLUSIONS

In the majority of cases where natural gas has to be transported from source to demand centre, the means of transport is dictated by geographical considerations.

A consideration of the potential characteristics of long-distance natural gas pipelines suggests there is a case for further attention to be given to the question of raising conventional operating pressures in pipelines to be designed in the future. Despite the cost of solving the appreciable engineering problems which would be raised by such a view, there appear at first sight to be substantial economic advantages in such a course.

## ACKNOWLEDGEMENTS

Thanks are due to the directors of CJB Ltd for permission to publish this paper. The opinions are those of the authors and do not necessarily represent the policy of the company.

## REFERENCES

1. *Anon, Oil Gas J.,* 3.8.70, 109-16
2. Hubbard, M. *J. Inst. Petrol.,* 1967, **53,** 1-21.
3. Katz, D. L., and Hashemi H. T. *Oil Gas J.,* 7.6.71, 55-60
4. Odell, P. R. *Petrol Times,* 11.2.72, 25-28.
5. Salkeld, J. *Hydrocarb. Process.,* April, 1971, 125-28.
6. Clar, R., Jones, D. M., and Owens, P. J. *J. Instn. Gas Engrs,* 1967, 7, 653-69.

## APPENDIX

### 1. Computer Programme

The core of the programme is the solution of the flow formula desçribed in 2, but the programme is initiated by the calculation of the working pressure of the pipe under question.

The pressure drop for an increment of length (one mile) is calculated for a range of inlet pressures working down from the maximum working pressure. The horsepower associated with these pressure drops is calculated by the method described in 4.

The horsepower data, together with the pipe characteristics, are then fed to the costing routine which estimates the capital costs, including construction, and carries on to calculate maintenance and operating costs, including fuel. The basis for the depreciation cost has been taken as a ten-year write-off period.

The unit cost of transporting the gas is then calculated from the above components.

### 2. Flow Formula

Ref:  Engineering Data Book,
Natural Gas Processors Suppliers
Association, 1966

$$Q = 115.1 \frac{Tb}{Pb} \left[ \frac{p_1^2 - p_2^2 - \left[ \frac{0.0375(h_2 - h_1)}{Zavg} \frac{p^2 \, avg}{Tavg} \right]}{G \quad \frac{Tavg}{} \quad \frac{Zavg}{} \quad L} \right]^{0.5} D^{2.5} \log_{10} \frac{3.7D}{Kc}$$

Q   =  Gas flow rate (ft $^3$/day at Pb and Tb)

Tb =  Absolute base temperature ($^O$R)

PB =  Absolute base pressure (psia)

$P_1$ =  Absolute initial pressure (psia)

$P_2$ =  Absolute final pressure (psia)

Kc =  Effective roughness of inside pipewall (inches)

Zavg =  Average compressibility factor at Pavg, Tavg

G  =  Gas specific gravity (air = 1.0)

$h_1$ =  Initial elevation (ft above sea level)

$h_2$ =  Final elevation (ft above sea level)

$T_1$ =  Initial temperature ($^O$F)

$T_2$ =  Final temperature ($^O$F)

D  =  Pipe inside diameter (inches)

L  =  Length (miles)

Pavg  =  $\frac{2}{3} \left[ P_1 + P_2 - \frac{P_1 P}{P_1 + P_2} \right]$   psia

Tavg  =  $\left[ \frac{T_1 + T_2}{2} \right] + 460$   but as

$T_1 = T_2$ for this analysis Tavg = $T_1 = T_2$

## 3. Working Pressure of Pipe

$$Pwork = \frac{2 \times \text{Pipewall thickness (inches)} \times \text{Working stress (psia)}}{\text{Diameter (inches)}}$$

The working stress has been taken as 80 per cent of the minimum yield strength.

## 4. Horsepower Calculations

Ref:   Engineering Data Book
       Natural gas Processors
       Suppliers Association, 1966.

Brake horsepower = $\dfrac{W \times H_p}{33000 \times E_p}$

W  =  weight flow lb/min.

Ep =  Polytropic efficiency (taken as 72.5 per cent)

Hp =  Polytropic Head

$\quad = \dfrac{1545}{\text{Mol. Wt.}} \dfrac{T \ Zavg}{M} \left( \left( \dfrac{P_2}{P_1} \right)^M - 1 \right)$

$P_1$  =  Intake pressure (psia)

$P_2$  =  Discharge pressure (psia)

Zavg  =  $\dfrac{Z_1 + Z_2}{2}$

$Z_1$ =  Compressibility factor at intake temperature and pressure

$Z_2$ =  Compressibility factor at discharge temperature and pressure

M  =  $\left( \dfrac{K - 1}{K} \right) \Big/ E_p$

K  =  $\dfrac{C_p}{Cv}$  of gas

## 5. Gas Compressibility

As the gas compressibility factor is of considerable importance in the high pressure transmission of gas, it has been estimated using the Beattie-Bridgeman equation of state from the gas critical properties.

Ref:   "Compressibility Factor by Computer", Jude T. Sommerfield and Gerald L. Perry, H.P, October 1968, 47 (10).

## 6. Costs and Source of Costs

(a)   Pipe Cost
       All pipe costs have been based on the prices given by the British Steel Corporation for Hot Finished Tubes for linepipe to API 5LX grade X 52.

(b) Pipe Installation Cost
This cost has been based on CJB's experience. The figures used being quite general and, therefore, referring to no country in particular.

(c) Compressor Station Cost
As no size or position of station is established, a general cost of £100.8 per installed horsepower has been used. This is based on the assumption that the compressors will be gas turbine-driven and the figure for cost is taken from the construction costs reported to the Federal Power Commission in America for the year 1970. The horsepower calculation has assumed a polytropic efficiency of 72.5 per cent.

(d) Pipeline and Compressor Station Maintenance and Operating Costs (excluding fuel)
The operating and maintenance cost of the pipeline and compressor stations have been taken as a total of 5 per cent of the capital cost of the two per year.

(e) Gas Turbine Fuel Cost
The cost of fuel for driving the gas turbines was taken as £12/HP - year; this is based on a loss of revenue from the sale of natural gas of 1.5p/therm (this is based on updated costs quoted in European Chemical News, Dec. 1968, p.6).

(f) Capital Costs
A ten-year "write off" period has been assumed, thus giving a cost of 10 per cent of the total capital cost per year. It must be noted that no allowance has been made for return on capital.

## DISCUSSION

*R. L. Torczon* (Conoco Europe) opened the discussion by stressing the importance of stabilizing the cost of energy and commodities if society was to maintain or improve its present standard of living. Natural gas, being in demand as a premium fuel, was being found in increasingly inaccessible places, and bringing this gas to market at an economic rate represented a major engineering challenge. Currently, offshore gas lines were limited to 1300 psi, and he would like the authors' views on the problems facing the adoption of high-pressure lines proposed in the paper. The inclusion of a rate of return on investment would have made the cost figures more meaningful. The load factor should be thoroughly evaluated, to minimize running expensive pipelines at below capacity. He suggested that an interesting subject for a future paper would be comparison of load factors for a long distance offshore pipeline (*e.g.* Frigg), as compared with peak shaving with LNG.

*Mr Stewart,* in reply to the points raised, said that the purpose of the paper was to provide a comparison of cost saving on high-pressure operation of 24- and 28-inch pipelines, and not to come up with realistic costs. Rates of return acceptable to various companies showed wide variation and he hoped the audience could provide some leads on them.

*Mr Roberts* added that rates of return on pipelines were a political problem akin to that of the price of gas. With regard to load factors, he considered there were too many dimensions to produce firm conclusions. The problems associated with operating high-pressure gas lines had been mentioned in the paper, and it was thought these could be overcome, with pipe manufacture likely to prove the most difficult technical problem.

*A. B. Densham* (The Gas Council) enquired if severe gas treatment problems would arise with transmission at 2000 psi with much more cooling required to eliminate heavier hydrocarbons. *Mr Roberts* replied that with transmission at 2000 psi retrograde condensation would not occur in the line, and the liquids could be removed at the end of the trunk line, prior to distribution to consumers.

*R. Evans* (The Gas Council) considered that high-pressure transmission lines would be excellent for uninhabited areas, but in urban areas they were not practical, due to risks of damage. In urban areas, damage by external interference was the main cause of failure. He also considered manufacture of X100 pipe could cause severe problems, as currently production of X60 pipe was proving difficult. The question of super-compressibility of high-pressure gas was also raised.

*Mr Stewart* replied that X100 pipe was not used in the cost comparison, but X52 and X60 with the increased wall thickness. *Mr Roberts* added that super-compressibility was not gone into in any detail, but considered peak shaving problems could be reduced by using Joule–Thomson effect of high-pressure gas for small-scale liquefaction.

# Elements in the Cost of Liquefaction of Natural Gas

By R. M. THOROGOOD, B. DAVEY
and W. HENDRY

*(Air Products Ltd)*

## LNG PERSPECTIVE

World consumption of liquefied natural gas is currently concentrated in Europe and Japan. The commissioning of the Skikda and Brunei plants (see Table I) will bring the total annual production capacity of LNG to $550 \times 10^9$ scf, of which a little less than half is destined for Japan, with the remainder in Europe. However, the European LNG market is unlikely to grow significantly further following the development of the indigenous gas fields. The major sales of LNG during the next decade are expected to occur in the U.S.A. accompanied by further growth of the Japanese market.

The first major contract to the U.S.A. will be, in all probability, from Sonatrach at Arzew, Algeria to the Atlantic seaboard at Cove Point, Maryland and Savannah, Georgia. This contract for an annual production of $520 \times 10^9$ scf/year will itself almost equal the previous total world production of LNG.

The supply gap between indigenous production and consumption of natural gas in the U.S.A. during the 1970s is expected to grow to $2500 \times 10^9$ scf/year[1]. The particular areas which can utilize LNG to ease the supply problem are the East Coast and Southern California. Possible imports to Southern California have been estimated as high as $350 \times 10^9$ scf/year and to the East Coast at $1500 \times 10^9$ scf/year by 1980.

Production, shipping, and reception of LNG in these large quantities require a large investment and can only be undertaken economically on a long-term supply contract basis. The magnitude of each individual contract is such as to warrant complex and detailed evaluation studies for the most economic process and equipment. Local conditions and the previous experience of the operating company must inevitably condition the detailed content and cost of a specific facility. It is not possible to generalize such specific situations and the following discussion of the general cost aspects should be considered in this context.

TABLE I

**Natural Gas Base Load Liquefaction Facilities — Existing and Planned**

| Liquefaction Plant | Receiving Terminal | Plant Owner | Contractors | Process | Annual Production Capacity scf. |
|---|---|---|---|---|---|
| Arzew, Algeria | Canvey Island, U.K. Le Havre, France | Camel | Technip - Pritchard | Classical cascade | $53 \times 10^9$ |
| Kenai, Alaska | Yokohama, Japan | Phillips — Marathon | | Classical cascade | $50 \times 10^9$ |
| Marsa-el-Brega Libya | Genoa, Italy Barcelona, Spain | Esso Standard Libya | Bechtel - Air Products | Mixed Refrigerant | $119 \times 10^9$ |
| Skikda, Algeria | Marseilles, France | Somalgaz | Technip - Air Liquide | Mixed Refrigerant | $150 \times 10^9$ |
| Brunei, Island of Borneo | Osaka and Tokyo, Japan | SIPM | Procon, Japan Gasoline, Bechtel, Air Products | Mixed Refrigerant | $200 \times 10^9$ |
| Skikda, Algeria | Marseilles, France | Somalgaz | Pritchard Rhodes | — | $50 \times 10^9$ |
| Arzew, Algeria | Savannah, Georgia Cove Point Maryland | Sonatrach | Preliminary Study Chemico- Air Products | Mixed Refrigerant | $520 \times 10^9$ |

In addition to the facilities listed, at least seven other schemes are known to be under construction, involving LNG production in the Middle East, Africa, and the Pacific for supply to the U.S.A. and Japan.

## LIQUEFACTION PROCESSES

It is significant that, of the base load liquefaction facilities currently operating or under construction, only the first two have similar process designs. These are the Kenai and Arzew plants, utilizing standard cascade cycles to liquefy 140 to $190 \times 10^6$ scfd of natural gas in three parallel processing lines. The remaining plants all use variants of the mixed re- frigerant cycle, which reflect the rapid development of the liquefaction process in a period of less than ten years. The various processes have been previously described in substantial detail[2-5]. A brief description only is given below to assist in illustrating the major economic parameters.

### The Standard Cascade Cycle

The standard cascade cycle for natural gas liquefaction (see Fig. 1) uses three separate refrigerants, namely, methane, ethylene, and propane, to cool and liquefy the natural gas. Each refrigerant requires separate compression and heat exchange equipment and hence a minimum of three separate closed cycles is required in a single liquefaction line. To achieve an economic power consumption, each refrigerant system will have several operating pressure levels, with each pressure level requiring its own heat exchange equipment.

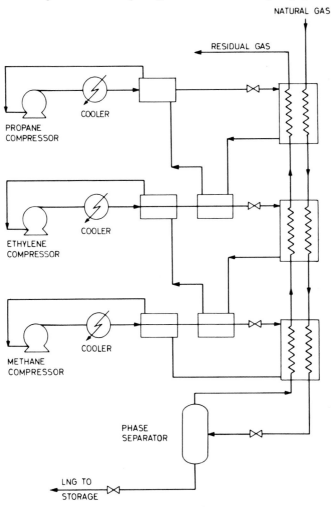

Fig. 1 Cascade cycle

It is generally accepted that the standard cascade cycle requires less power than the simple mixed refrigerant cycle. However, current developments of the mixed refrigerant cycle have virtually eliminated this power differential, and it is clear from evaluations of the capital cost that the standard cascade cycle is no longer competitive.

The significant differences in capital cost have been demonstrated by Mellen and Pryor[2] and Bourguet et al[8]. The cost difference due to compression and drive equipment, liquefaction process equipment (including piping and instruments) is shown to be approximately 20 per cent in favour of the mixed refrigerant cycle, resulting in a total cost reduction of up to £8 x $10^6$ for a plant liquefying 200 x $10^6$ scfd of natural gas. This is approximately equal to the total fuel cost of operation over a 20-year period and hence cannot be justified by any conceivable power saving.

## The Mixed Refrigerant Cycle

The remote location of the majority of base load plants imposes a particular consideration in the selection of the process cycle, namely, that there will be a shortage of skilled labour at the site, both for construction and for subsequent operation.

In consequence, because of the difficulty and cost of site construction, it is extremely desirable to utilize pre-assembled and packaged components, but with the minimum possible number of separate units. Similarly, operation of the plant is simplified by reducing the number of liquefaction lines to a minimum, together with the number of equipment items in each line.

These factors, coupled with the low cost of energy at the gas source, have provided a strong incentive for the development of the mixed refrigerant process with its complex thermodynamic and engineering design features.

The simple mixed refrigerant cycle is illustrated in Fig.2. In its basic form a single refrigerant is used which consists of a mixture of light paraffins (from $C_1$ through $C_5$) with nitrogen. This refrigerant is compressed and partially condensed against cooling water or air. The condensed liquid fraction is expanded to the low pressure of the cycle and is used as a first stage of refrigeration to cool the natural gas and to further partially condense the remaining refrigerant. The process of partial condensation and phase separation of the refrigerant is repeated as many times as necessary to achieve an economic balance between the power consumption of the process, the capital cost of the compression system, and the capital cost of the heat exchangers. The expanded liquid fractions are successively re-combined at each stage of refrigeration and are re-cycled through the refrigerant compressor.

It is the practice in a base load installation to generate the principal components of the refrigerant from the natural gas feed. The simplest method by which this may be achieved is to use an open cycle of the type

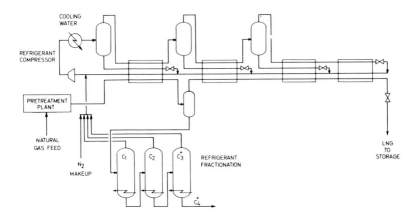

Fig.2 Mixed refrigerant cycle applied to a base load LNG plant

originally proposed by Kleemenko [9] and illustrated in Fig.3. However, this cycle suffers from the restraint placed upon the composition of the refrigerant by the incorporation of the natural gas. A closed refrigerant cycle is used in preference, with the hydrocarbon components being recovered from the natural gas feed by fractionation and with nitrogen supplied from an external generator. This allows the required flexibility to optimize the plant operation with varying feed gases.

The major advantages of the simple mixed refrigerant cycle may be stated as:

Minimum capital investment
Self provision of the principal refrigerants
Operational flexibility to handle varying feed gases.

Variations from the Simple Mixed Refrigerant Cycle

The simple mixed refrigerant cycle utilizing a single compression train and heat exchanger installation can accommodate a liquefaction capacity of approximately $120 \times 10^6$ scfd. To achieve this, an installed compression power of approximately 65 MW would be required which would consume from 15 to 18 per cent of the feed gas as fuel. The principal limitation to this capacity is set by the shipping dimensions of the wound coil heat exchangers. It also fits conveniently the use of two centrifugal compressor casings in series with each close to its maximum power rating.

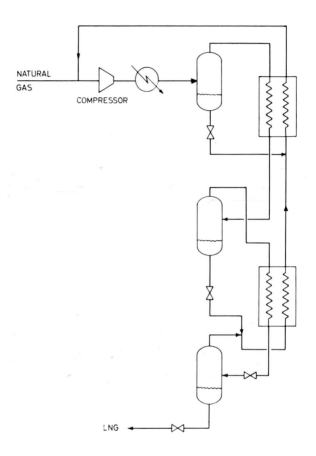

Fig. 3 One flow cascade cycle of Kleemenko

In order to achieve a larger liquefaction line capacity, it is necessary to use more than one heat exchanger assembly. This accepted, it is then sensible to consider a more complex cycle design in which the power consumption may be reduced. The power disadvantage of the mixed refrigerant cycle when compared with the standard cascade results in part from the entropy loss of the continuous separation and re-mixing of components. A power reduction may thus be achieved as in the cascade cycle by using two or more refrigerant circuits, e.g. by using a stage of propane pre-cooling, followed by a mixed refrigerant cycle operating between, say, -30º and - 270ºF.

Alternative schemes, in which a power reduction is achieved by the use of an additional pressure level, for the refrigerant, have been described by Bourguet et al.[8] and in a U.S. Patent.[10] The latter scheme utilizes partially condensed refrigerant from an intermediate compression stage as a high-level refrigerant for pre-cooling of the natural gas and the high pressure refrigerant. The scheme described by Bourguet separates the functions of refrigerant condensation, and natural gas cooling and condensation, with an intermediate evaporation pressure being used for cooling of the high pressure refrigerant. This is shown in Fig.4.

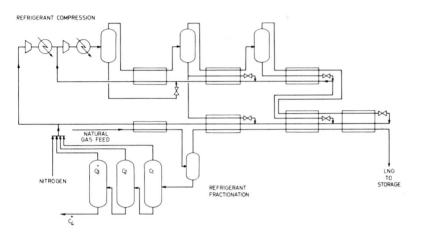

Fig. 4 Dual pressure mixed refrigerant cycle

The use of these and other more complex variations permits the liquefaction of greater than $150 \times 10^6$ scfd in a single production line, and allows the utilization of axial compressors.

In considering the use of production lines of such large capacity, it is of extreme importance to obtain a very high degree of equipment reliability and simplicity of operation in order to minimize production losses. Both of these factors give preference to the use of equipment with a proven reliability and have, for example, favoured the use of steam turbine drivers for compression. The requirement for reliability goes in parallel with the conditions for safety which must accompany the handling of large quantities of volatile liquefied hydrocarbons.

A particular requirement in this respect is a high integrity of the equipment in those parts of the process where liquid is present, namely, the heat exchangers and process piping. Large coil-wound heat exchangers offer certain inherent advantages in that the extensive high pressure heat transfer surface is contained within a small number of medium pressure shells which would act as a secondary barrier in the event of a tube rupture. In addition, the number of pipe circuits is minimized, thus reducing the probability of leaks.

## Peak Shaving Plants

The majority of facilities for LNG peak shaving require a liquefaction capacity in the range of 5 to 25 x $10^6$ scfd. A number of special features distinguish the peak shaving plant from the larger base load plant.

Occasionally, the plant may be associated with a gas distribution centre at which a substantial gas pressure let-down may occur. In this case, the use of an expander cycle (as, for example, described by Markbreiter and Weiss [11] allows the inherent generation of sufficient refrigeration to liquefy up to 10 to 15 per cent of the total gas flow.

The peak shaving plant is generally located in industrial areas where refrigerants are obtainable economically as bulk liquid supplies. A considerable freedom is thus available to select refrigerants such as ethylene and propylene in the process cycle.

The cost of energy becomes a determining factor in the selection of the process cycle. This is illustrated by Mellen and Pryor[2] in a comparison of a standard cascade, a mixed refrigerant cascade, and a nitrogen re-cycle system using an expansion turbine. In this, comparison it was shown that at a natural gas cost of 12p per million Btu compared to 0.46p per kilowatt hr for electricity, the use of gas engine drive would be favoured and the lower power and capital cost of the mixed refrigerant cycle would determine its choice. However, reduction of electricity cost to 0.19p per kwh would favour the nitrogen cycle.

## PROCESS EQUIPMENT

The major items of process equipment in the liquefaction plant comprise:
Natural gas pre-treatment
Heat exchangers and associated process equipment
Refrigeration compression and drive systems
LNG storage
These items constitute from 50 to 80 per cent of the total capital investment. The remaining investment is required for site development, off-site utilities, including cooling and disposal systems, spares, personnel requirements, etc. An approximate breakdown of costs is given in "Summary of Liquefaction Plant Costs" (p. 92).

## Natural Gas Pre-treatment

Prior to liquefaction of gas, it is necessary to remove all impurities which would freeze out and plug the heat exchanger equipment. Assuming that the gas has previously been freed of condensate at the gas field or in a condensate removal plant, then it may be necessary to remove water (1 ppm), carbon dioxide (150 ppm), hydrogen sulphide (700 ppm), and trace aromatics, *e.g.* benzene (2 ppm). The approximate permissible concentrations of each are given in parentheses.

Removal of $CO_2$ and $H_2S$ is most commonly accomplished by continuous absorption processes, especially for $CO_2$ concentrations in excess of 0.5 per cent. Selection of the appropriate absorption process permits direct achievement of the required acid gas concentrations.

Drying, because of the low dewpoint required, can only be accomplished by solid bed adsorption. A choice exists between alumina and molecular sieves, the lower costs of the former being offset by the higher capacity and hence lower vessel and reactivation heating costs of the latter. The cost of either is reduced by operation at the minimum possible temperature. This temperature is limited by the hydrate formation temperature of the gas, which may be as high as 65°F. It is necessary in the evaluation of adsorbents to take considerable care that undesirable side effects are not introduced, such as the accumulation and subsequent release of slugs or other components of the natural gas.

Removal of aromatic components, *e.g.* benzene, is a more difficult problem, since the acceptable level of benzene in the feed gas is a function of the quantity of the other heavy hydrocarbons, especially those which condense from the natural gas at temperatures from ambient down to -60°F. In some particular situations, aromatics may be removed satisfactorily, with the heavy paraffins being recovered for refrigerant purposes. Alternatively, it may be necessary to scrub the feed gas with a condensed fraction of the natural gas.

## Heat Transfer Equipment

Three types of heat exchanger are more applicable for natural gas liquefaction:

Shell and tube

Brazed aluminium plate-fin

Wound coil

Shell and tube exchangers are primarily suited to the warmer (propane) level of refrigeration, at which it is not important to achieve very close temperature approaches. The recent introduction of very high efficiency boiling surfaces may extend the range of application to lower temperatures, although the present cost of such surfaces very largely offsets their advantages.

The brazed aluminium plate-fin exchanger is used extensively in cryogenic plants. Its use in base load liquefaction is limited by the maximum size of units which are available for the required operating pressure of 550-700 psi. For example, a $100 \times 10^6$ scfd plant would require up to 25 exchangers in parallel at the warm and intermediate temperature zones. Apart from the cost of manifolding the exchangers, it is necessary to consider the feasibility of obtaining stable two-phase flow conditions for simultaneous condensing and boiling of at least two streams in such an array. Whilst the use of this type of exchanger is well established in pool boiling and thermosyphon applications such as exist in standard cascade plants, they have not yet been proven in a base load mixed refrigerant plant.

The economic solution adopted to meet the larger surface area requirements of base load plants has been the coil-wound aluminium heat exchanger. The current maximum capacity of a liquefaction plant line is determined primarily by the capacity of the heat exchangers. The heat exchanger capacity is in turn limited by the maximum dimensions of an exchanger which can be fabricated and shipped. Coil-wound exchangers can at present be manufactured with dimensions in excess of 12 ft and with bundle lengths in excess of 50 ft. The final shell assemblies for the Esso Libyan plant heat exchangers were 200 ft in length and were shipped complete from the U.S.A. to Libya. Each complete shell assembly contained in excess of 250,000 $ft^2$ of surface to liquefy over $90 \times 10^6$ scfd of gas.

### Compressors and Drivers

The capital investment in the compression and drive system constitutes from 30 to 40 per cent of the total liquefaction plant investment (see cost summary). It is obvious that the selection of equipment must be made with extreme care. Significant reductions in the operating cost can be made by optimizing the efficiency of both the cycle and the compression system, but of even greater importance is the reduction of capital investment that can result from a proper combination of equipment.

The principal features to be considered for the compression system on a base load plant are:

The overall compression ratio of the MCR compressor will be approximately 12 for a typical cycle

The installed power requirement for each liquefaction line will be from 50 to 150 MW

A selection must be made between axial and centrifugal compressors from a consideration of their relevant characteristics

A choice exists between steam and gas turbine drive

### Axial — Centrifugal

In a comparison of axial with centrifugal compressors for large duties, several factors are immediately apparent.

At the same power rating, the operating speed of a centrifugal compressor is approximately two-thirds of the speed for an axial compressor.

A typical axial compressor for LNG service approaches the speed of synchronous turbines (3000 and 3600 Mr/min) at an approximate power rating of 30 MW. At power levels in excess of 30 MW, the operating speed reduces progressively (speed is inversely proportional to the square root of power) and the axial compressor is preferred to the centrifugal as being more readily matched to the speed of the turbine drive system.

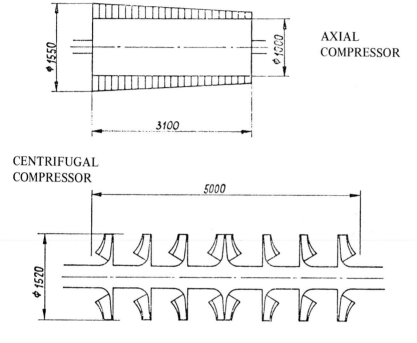

Fig. 5 Comparison of the dimensions of similar duty axial and centrifugal compressors
(Courtesy of Brown Boveri — Sulzer Turbo Machinery Ltd, Zurich)

The centrifugal compressor stage is equivalent to approximately three axial compressor stages. A 12-stage axial may be compared with an eight-stage (two x four stages) parallel flow centrifugal (see Fig.5) in which, due to the halving of inlet flow, the rotor diameters will be the same for the same power demand, say, 50 MW at 3000 Mr/min. (This arrangement of centrifugal compressor is known as back to back). However, the rotor length will be 60 per cent greater for the centrifugal.

The second critical speed of the centrifugal compressor is lower than that of the axial.

The axial machine has only two pipe nozzles compared with three on the centrifugal.

The centrifugal has a longer, and hence heavier, more expensive casing.

### Effect of Suction Pressure

Should the suction pressure be raised above 2.6 bar to, say, 4.0 with a compression ratio of 6, and the intake volume kept the same (hence doubling the power), then the centrifugal compressor may still be competitive at 40/60 MW, as the axial would run too fast for existing drivers.

### Compressor Problems

A difficulty with the axial compressor is its susceptibility to blade failures. However, the manufacturers are now confident that this problem has been resolved. Compressors are now under construction which have a power input of 80 MW into two casings.

It is not usually convenient to combine axial and centrifugal compression stages in the same compressor because of the required speed differential for equal power distribution. A speed increase of 37 per cent would be required on the centrifugal stage to obtain even distribution of power and of compression ratio. However, the use of gear boxes for power transmission on the scale required is undesirable and acts as a strong disincentive for the hybrid compressor.

### Drivers

There are two types of driver which can be considered for application with either axial or centrifugal compressors, namely, gas turbine or steam turbine.

The technological advances in this area have been primarily around the steam driver which, for historical reasons, has resulted in a more robust machine having acceptance throughout the industry for its reliability. Steam turbines with power outputs of up to 200 MW are designed and up to 150 MW have been built. For the gas turbine, maximum power capacities of 100 MW have been developed but only 50/50 MW turbines have been built.

It is thus apparent that with increasing size of base load LNG plant liquefaction lines, any attraction which the gas turbine may have had is rapidly being lost to the steam driver.

For the size of plant now being considered, it is essential to achieve a very high on-stream factor and hence complete dependability of every component. As a result, only tried and proven designs can be accepted and extrapolations of the technology are permitted only when very strong economic incentives exist. It may be desirable in some instances to utilize schemes whereby, in the event of failure of a component (*e.g.* compressor, driver, or heat exchanger), it is possible to interconnect equipment and thus minimize the effect of individual failures. The differing liquefaction processes may vary in their ability to permit operation at reduced load capacities and this is a factor to consider in selection of the compression system.

## LNG Storage

Both within Europe and the U.S.A. it is common to use above-ground double-walled metal tanks for storing LNG, although it must be pointed out that a tendency now exists in some European countries, notably Germany, to insist on the outer tank being capable of retaining the inner tank contents in the event of a rupture. This recent regulation has necessitated the design and construction of three-walled vessels. This is a consequence of the need to minimize the vapour release which would result from the inner tank contents contacting a warm outer shell as in the case of a double-walled vessel.

One alternative to above-ground storage is in-ground storage, where the earth itself is frozen by the LNG and then used as the insulating medium to protect against boil-off. A number of such installations have been built, of which the more successful have been associated with base load plants, *e.g.* Canvey receiving terminal [12]. Experience has shown, however, that the boil-off losses occurring in practice have been in excess of the predicted values. The boil-off loss from a typical double-walled tank amounts to 0.1 per cent of tank contents per day, whilst that for in-ground storage is approximately 0.5 per cent (predicted) and can be up to 1.0 per cent in practice. For peak shaving installations, at least, these losses reverse the economic decision to provide in-ground rather than the double-walled "conventional" tanks.

Any natural gas evaporated at low pressure at the receiving terminal must be compressed to the distribution pressure, resulting in additional capital expenditure and an increase of operating cost for in-ground storage when compared to a double-walled tank. Additionally, boil-off from a "heavy" LNG, *e.g.* Algerian gas landed at Canvey, will result in the residual liquid "heavying-up". This can be serious enough to affect the calorific value and Wobbe number, thus requiring additional equipment to modify the gas composition during subsequent vaporization.

Equipment to avoid this problem is being installed at Canvey Island, where Air Products are constructing a 200 LT/D North Sea Gas Liquefier, which is able to act as a boil-off re-liquefier. Included in the contract is an additional exchanger capable of re-liquefying approx. 200 LT/D of compressed boil-off against "send-out" liquid. By using this facility, the storage liquid composition can be retained at a constant level.

Two types of pre-stressed concrete container have also been developed. One design utilizes insulation on the outside of the concrete, with the concrete at cryogenic temperature, whilst the other uses internal insulation. Both designs use an internal liner to ensure a liquid-tight seal.

This type of tank can be used either below ground or in a partially buried and bermed configuration. This avoids any requirement for diking to retain spills.

The concrete tank has only been found to be economic at sizes larger than 600,000 brl on the East Coast of the U.S.A.

## THE INFLUENCE OF NATURAL GAS COMPOSITION IN LIQUEFACTION

Throughout the foregoing, the basic elements of the liquefaction plant have been considered without specific reference to the composition of the feed gas. The variation of composition encountered between gas fields and even between wells in the same field is quite substantial. The basic constituents of natural gas are listed in Table II, together with an indication of the range of concentrations which may be encountered. In addition to these principal components, a large number of trace components occure, which on occasion may introduce processing problems, *e.g.* trace quantities of heavy hydrocarbons may induce a high dewpoint temperature and introduce unexpected two-phase flow conditions.

Water and carbon dioxide removal have already been discussed and will not be treated further.

### Hydrocarbons

Variations in the content of ethane and heavier hydrocarbons influence the calorific value of the gas and it may be necessary to substantially modify the composition of LNG to achieve compatibility with existing gas distribution networks at the receiving terminal. A heavy gas such as occurs in North Africa may need to be diluted with an inert gas, or the gas may be processed for removal of propane, butane, etc., either at the liquefaction plant or the receiving terminal. Recovery of these hydrocarbons in the liquefaction plant causes an increase in the power requirement for liquefaction of the lighter residual gas. A power increase of 8 to 12 per cent has been quoted [7] for removal of 30 per cent of the ethane and all of the propane and heavier components from an Algerian gas.

Whilst a limited recovery of hydrocarbons is necessary at the liquefaction plant for use as refrigerant, the most economic route for their recovery is likely to be during re-evaporation at the receiving terminal, when they will become available for marketing as LPG in the surrounding area.

**Aromatic and Organic Sulphur Compounds**

Benzene, toluene, and other aromatics present a problem of having very low solubilities in LNG. Benzene, in particular, has a low solubility (of the order of 2 ppm) and is known to be present at concentrations of up to 500 ppm in North Sea gas, where it presents a difficult processing problem. Organic sulphur compounds, particularly di-ethyl sulphide, have been found at low ppm concentrations in North Sea gas and may occur in other gases. These also have low solubilities and hence may constitute a problem. They have a tendency to be adsorbed in solid-bed pre-treatment systems and in the case of peak shaving plants may constitute a nuisance by causing peaks in the odourant content of reactivation gas being returned to the distribution main.

### TABLE II
### Typical Natural Gas Compositions

| Component | Algerian gas Mole % | North Sea gas Mole % | Extreme Composition Range of natural gases Mole % | |
|---|---|---|---|---|
| $CH_4$ | 83.6 | 93.1 | 5.1 | 98.0 |
| $C_2H_6$ | 7.0 | 3.2 | 0 | 71.3 |
| $C_3H_8$ | 2.1 | 0.6 | 0 | 20.5 |
| $C_4H_{10}$ | 0.9 | 0.2 | 0 | 8.2 |
| $C_5H_{12}$ | 0.2 | 0.1 | 0 | 3.2 |
| $C_6H_{14}+$ | 0.1 | 0.2 | 0 | 1.7 |
| $N_2$ | 5.7 | 2.0 | 0 | 84.5 |
| $CO_2$ | 0.2 | 0.5 | 0 | 29.0 |
| He | 0.2 | 0.1 | 0 | 7.8 |
| $H_2S$ | – | – | 0 | 5.2 |
| $O_2$ | – | – | 0 | 0.5 |

In peak shaving plant of $10 \times 10^6$ scfd capacity, equipment for treatment of aromatics and organic sulphur compounds may constitute 10 per cent of the capital cost of the liquefaction plant.

**Nitrogen**

Nitrogen acts to reduce the calorific value of the natural gas and also to reduce the LNG temperature, thus increasing the power required for liquefaction and the cost of storage. For example, an increase from a nitrogen content of 1 to 3 per cent in the LNG may cause a power increase for liquefaction of 13 per cent.

High nitrogen contents are encountered in a number of natural gases, especially Dutch gas (14 per cent) and some Algerian gas (6 per cent).

### Helium

Most natural gas resources being considered for liquefaction do not contain helium at a sufficient concentration in the raw gas to warrant its direct extraction. However, helium accumulates in the flash gas and boil-off from storage to appreciable concentrations (5% or greater). Hence an opportunity may exist to recover helium in large quantities at a reasonable concentration for processing at the liquefaction site.

## SUMMARY OF LIQUEFACTION PLANT COSTS

In this section, estimates of the capital and operating costs are presented for two base load facilities having capacities of $500 \times 10^6$ scfd and $1000 \times 10^6$ scfd respectively. In order to obtain an overall picture of the cost of LNG when re-converted to a usable gas form at the final point of distribution, estimates of the cost of shipping and of receiving terminals have been included.

### Liquefaction

The capital and operating costs for a base load liquefaction plant are given below.

| Liquefaction plant size scfd x $10^6$ | 500 | 1000 |
|---|---|---|
| Capital £ x $10^6$ | 90.0 | 150.0 |
| Annual operating expenses £ x $10^6$ | | |
|     Internal gas consumption | 1.1 | 2.1 |
|     Personnel | 1.4 | 2.3 |
|     Maintenance | 1.6 | 2.8 |
|     Expendables, miscellaneous | 0.2 | 0.4 |
|     Insurance | 0.6 | 1.1 |
| | 4.9 | 8.7 |
| Annual fixed charges          £ x $10^6$ | | |
|     Depreciation  )<br>    Interest        )<br>    Taxes          )          at 18.5%<br>    Overheads    )<br>    Profit          ) | 16.7 | 27.8 |
| | 21.6 | 36.5 |

| | | |
|---|---|---|
| Annual volume shipped scf x $10^9$ | 170 | 340 |
| Added value by LNG plant p/1000 scf | 12.7 | 10.7 |
| Range of values of feed gas p/1000 scf | 2–6 | |
| Feed gas value used p/1000 scf | 4 | |
| LNG delivered to ship p/1000 scf | 16.7 | 14.7 |

The fixed charges indicated above are only typical for natural gas pipeline companies using 25-year asset lives, and will vary depending on debt cost, asset life used, risk assessment on the owner's part, etc. The important point emphasized is the high capital intensity involved. The resulting heavy fixed charges are substantially higher than the operating costs.

Typically, liquefaction facility costs fall into the following approximate ranges:

| | % |
|---|---|
| Site development | 4–6 |
| Feed gas treatment | 3–5 |
| Liquefaction and fractionation | 20–24 |
| Compression (feed and refrigerant) | 14–18 |
| Steam and power generator and distribution | 17–21 |
| Storage, loading, mooring | 12–15 |
| Offsites, including cooling systems, refrigerant make-up, vent and flare system, maintenance facilities, employee amenities, etc. | 13–16 |
| Start-up, spares, training programme, land | 5–7 |
| | 100 |

Substantial variations in LNG plant costs are due to differing feed gas compositions, LNG composition, plant location, mechanical equipment selection, local construction costs, and, finally, the time period in which the facilities are installed.

**Shipping**

Typical economics based on large tankers are as follows:

| | | | |
|---|---|---|---|
| LNG cargo capacity brl x $10^3$ | 750 | 460 | 460 |
| Capital investment £ x $10^6$ | 30.0 | 22.0 | 22.0 |
| Voyage length, (statute miles) | 5000 | 5000 | 2500 |
| Approximate round trips/year | 17 | 17 | 30 |

Annual expenses £ x 10⁶

| | | | | |
|---|---|---|---|---|
| Operating costs* | | 0.6 | 0.5 | 0.5 |
| Port charges | | 0.1 | 0.1 | 0.2 |
| Fuel (LNG boil-off)† | | 0.1 | 0.1 | 0.1 |
| Fuel (oil) | | 0.3 | 0.3 | 0.3 |
| | Total | 1.1 | 1.0 | 1.1 |

Annual fixed charges £ x 10⁶

| | | | | |
|---|---|---|---|---|
| Depreciation ) | | | | |
| Interest ) | | | | |
| Taxes ) | at 18.5% | 5.6 | 4.1 | 4.1 |
| Overheads ) | | | | |
| Profit ) | | | | |
| | Total | 6.7 | 5.1 | 5.1 |

| | | | |
|---|---|---|---|
| Value added to LNG | 14.5 | 18.2 | 10.5 |
| Equivalent to p/1000 scf/100 miles | 0.29 | 0.36 | 0.42 |

This value per 100 miles compares with natural gas pipelining costs (based on new construction) of 0.4 to 0.8 per 1000 scf per 100 miles.

The 18.5 per cent fixed charge is only typical but it is again evident that the fixed charges are substantially higher than operating expenses.

*   Includes foreign crew cost, stores, insurance maintenance and repairs, manning agent, and shore administrative expenses.

†   LNG heat leak is usually sufficient to provide 75-80 per cent of the ship fuel, with the balance from Bunker C fuel. For return voyages a "heel" of 2½ to 3 per cent of the tank contents is retained on board, to keep the tanks at their design temperatures of -285°F, and to provide fuel for the return voyage.

### Receiving Facilities

LNG receiving facilities vary widely depending on many factors, but will all generally include:

(a)   Dock facilities for the tanker size to be serviced.

(b)   Unloading facilities for ship unloading in 10-15 hours

(c)   Storage facilities for at least one ship load and probably two or more.

(d)   Pumping and vaporization facilities to delivery maximum peak flows at pipeline pressure.

(e)   Connection to mainline system.

(f)   Possible LPG recovery system.

Economics of two receiving terminals:

| | | |
|---|---|---|
| Storage volume brl x $10^3$ | 1500 | 900 |
| Send out scfd x $10^6$ | 500 | 300 |
| Capital investment £ x $10^6$ | 22.0 | 16.0 |

Annual operating expenses £ x $10^6$

| | | |
|---|---|---|
| Fuel and power | 0.4 | 0.2 |
| Personnel | 0.1 | 0.1 |
| Maintenance | 0.1 | 0.1 |
| Insurance | 0.2 | 0.1 |
| Total | 0.8 | 0.5 |

Annual fixed charges £ x $10^6$

| | | | |
|---|---|---|---|
| Depreciation ) | | | |
| Interest ) | | | |
| Taxes ) | at 20% | 4.4 | 3.2 |
| Overheads ) | | | |
| Profit ) | | | |
| Total | | 5.2 | 3.7 |

| | | |
|---|---|---|
| Annual throughput scf x $10^9$ | 180 | 110 |
| Value added to LNG p/1000 scf | 2.9 | 3.4 |

In this case 20 per cent has been used for annual fixed charges to account for rates. Again the high capital intensity is emphasized.

**Conclusion**

The elements of investment, capital charges, and costs required to execute a "typical" LNG project are summarized. No LNG project can be called typical, due to the many factors which can influence the economics involved. The summarized total cost position is based on units of 500 and 1000 x $10^6$ scf/day. These quantities are "delivered to pipeline" at the receiving end and thus are net of any intermediate losses.

| | | |
|---|---|---|
| Capacity scfd x $10^6$ | 500 | 1000 |
| Shipping distance (statute miles) | 2500 | 5000 |
| Ship size brl x $10^3$ | 460 | 750 |
| Investment of £ x $10^6$ | | |
| Liquefaction plant and terminal | 90 | 150 |
| Ships | 88 | 240 |
| Receiving terminal(s) | 22 | 44 |
| Total | 200 | 434 |

Operating expense £ x $10^6$/Y (p/1000 scf)

| | | | | |
|---|---|---|---|---|
| Liquefaction plant and terminal | 4.9 | ( 2.7) | 8.7 | ( 2.4) |
| Ships | 4.4 | ( 2.4) | 8.8 | ( 2.4 |
| Receiving terminals | 0.8 | ( 0.4) | 1.6 | ( 0.4) |
| Total | 10.1 | ( 5.5) | 19.1 | ( 5.2) |

Fixed charges £ x $10^6$/Y (p/1000 scf)

| | | | | |
|---|---|---|---|---|
| Liquefaction plant and terminal | 16.7 | ( 9.2) . | 27.8 | ( 7.6) |
| Ships | 16.4 | ( 9.0) | 44.8 | (12.3) |
| Receiving terminals | 4.4 | ( 2.4) | 8.8 | ( 2.4) |
| Total | 37.5 | (20.6) | 81.4 | (22.3) |
| Grand total | 47.6 | | 99.5 | |
| Value added p/1000 scf | 26.1 | | 27.5 | |
| Gas at plant p/1000 scf | 4.0 | | 4.0 | |

Total value of gas into pipeline p/1000 scf  30.1         31.5

The Btu value of LNG can vary from as low as 1000 Btu/scf to 1300 – 1400 Btu/scf. A typical value of 1100 Btu/scf has been assumed. The comparative value on a Btu basis of the above pipeline gas would be:

Total value of gas in pipeline p/Therm HHV  2.7        2.9

Another way to comprehend the total picture is to group together fixed charges and the operating expense element of each project part:

Costs £ x $10^6$/year (p/1000 scf)

| Capacity scfd x $10^6$ | 500 | | 1000 | |
|---|---|---|---|---|
| Gas at plant | 7.3 | ( 4.0) | 14.6 | ( 4.0) |
| Liquefaction plant and terminal | 21.6 | (11.8) | 36.5 | (10.0) |
| Ships | 20.8 | (11.4) | 53.6 | (14.7) |
| Receiving terminals | 5.2 | ( 2.9) | 10.4 | ( 2.8) |
| Total 4 x $10^6$/year | 54.9 | | 115.1 | |

As discussed previously, the costs for actual projects can vary from the above, depending on many factors. Particularly sensitive are the investment costs for liquefaction and receiving terminals, the gas price delivered to the liquefaction plants, and the ocean shipping distance. Many local factors can have a significant influence on these costs. In general, however, it is felt that the costs shown are representative and reasonable on an overall basis.

## ACKNOWLEDGEMENTS

The authors would like to thank the directors of Air Products for permission to publish this paper. Our particular gratitude is due to staff of Air Products and Chemicals Inc. for their assistance in preparation of cost data.

## REFERENCES

1. Mellen, A.W., and Litwak M. "LNG Perspectives and Economics". Society of Petroleum Engineers (AIME) Symposium on Petroleum Economics and Evaluation, 9 March 1971.
2. Mellen, A. W., and Pryor, J. A. "Liquefaction Cycles for LNG" 11th International Gas Conference, Moscow 13 June 1970.
3. Thorogood, R. M. "Mixed Refrigerant Process for Natural Gas Liquefaction". Institute of Refrigeration, London, 2 December 1971.
4. Culbertson, W. L. and Horn, J. "The Phillips Marathon Alaska to Japan LNG Project". 1st International Conference on LNG, Chicago, 7-12 April 1968.
5. Salama, C., and Eyre, D. V., *Chem. Engng. Progr.* 1967, **63**, 62-67.
6. Pierot, M. "Operating Experiences of the Arzew Plant". 1st International Conference on LNG, Chicago, 7-12 April 1968.
7. Bourguet, J. M., "Liquefaction of Natural Gas, Selection of Refrigerating Compression Drivers for Large Capacity GNL Units". 1st International Conference on LNG, Chicago, 7-12 April 1968.
8. Bourguet, J. M., Garnaud, R., and Grenier, M. *Oil Gas J.,* 30.8.71, 71-75.
9. Kleemenko, A. P. Proceedings International Congress of Refrigeration 1, 34-39, 1959.
10. U.S. Patent 3, 364, 685.
11. Markbreiter, S. J., and Weiss, I. *Cryogenic Engng News,* March 1968.
12. Ward, J., and Egan, P. Proceedings of the International Conference on Liquefied Natural Gas, London, March 1969.
13. Private communication from Brown Boveri — Sulzer Turbo Machinery Ltd, Zurich.

## DISCUSSION

The paper was presented by *B. Davey.* After setting world consumption of LNG in perspective, pointing out that LNG demand was currently concentrated in Europe and Japan, and touching briefly on the outlook in the U.S.A., Mr Davey dealt with the problem of the size of the equipment and the related economics.

*Dr M. J. Stacey* (British Oxygen Co. Ltd) raised two points. He said that Mr Davey had referred to the necessity for multi-train plant designs, which was a direct consequence of limits to sizes of certain major plant items. Such large units as the coil-wound heat exchangers used in the Esso Libya liquefaction plant clearly presented considerable fabrication problems, and Dr Stacey queried the relative merits of building and shipping over long distances such large units, instead of using smaller,

more widely obtainable units. He also invited comment on the thermal
design methods for those complex types of exchanger. To what extent did
the authors consider that the very large exchanger represented "too
many eggs in a simple basket", and in the future would perhaps larger
plants make use of multiple, if somewhat smaller, heat exchange units.
As a final point, Dr Stacey raised the question of the removal of traces
of aromatics, which was well known to cause problems in cold parts of
the plant.

*Mr Davey* commented that with regard to size of equipment Air
Products took the view that it was better to go for large units than
smaller ones, with inherent.problems of valving, etc. Costs of larger units
were absorbed in the company. Having constructed the larger exchangers,
experience had shown exceptionally speedy development of these
exchangers.

With reference to the extraction of aromatics, Mr Davey referred to
the need to provide components for the mixed refrigerant and to the high
paraffinic contents. These components were removed by pre-cooling.
Scrubbing media were also used for removal of aromatics. With regard
to the higher paraffins, a detailed examination waś being made of the
costs of production of LNG using a mixed refrigerant system.

*S. E. Churchfield* (Burmah Oil Trading Ltd) referred to the large capital
requirements and questioned whether estimates quoted in the paper
related to "today's or tomorrow's costs". He asked what assumptions had
been made as to the rate of inflation. *Mr Davey* replied that the figures
were in terms of "today's costs" and related to experience. As to inflation,
current trends indicated a rate of 10 per cent/year over the next four to
five years, which was the contract period currently being envisaged.

*G. L. Fabry* commented that from the presentation it appeared that
the distribution problem still existed for the two main types of
exchangers, *i.e.* the plate/fin type and the wound cross-flow type. This
problem had been successfully solved some time ago with the plate/fin
type exchangers as demonstrated in the existing Glenmavis LNG plant and
represented a solution where the scaling up to a much larger size exchanger
did not magnify the problem. Mr Fabry further referred to the paper's
forecast of future single trains with a drive of 150/200 MW and asked
whether Air Products intended to increase the size of exchangers to meet
this requirement. *Mr Davey* replied that development work was in hand but
the matter was of a confidential nature and he could not enlarge.

*J. M. C. Bishop* (Phillips Petroleum Co. Ltd) asked what was the overall
efficiency factor in terms of the percentage of gas available ex-shore plant,
compared with inlet crude gas to the liquefaction plant. *Mr Davey* replied
that the percentage varied with the quality of gas available. In general,
with good quality gas between 90 and 93 per cent should be available.

# LNG Carriers – the New Liners

By J. M. SOESAN and R. C. FFOOKS

*(Conch Methane Services Ltd)*

The transport of liquefied natural gas at sea involves the construction of special and high cost carriers which are generally purpose-built for each project. The designs of such LNG carriers currently available or being offered are briefly classified and reviewed and the factors affecting the price of the various designs examined. The effect of unscheduled outage time of LNG carriers on project economics is dealt with, since this is a factor in design choice. Further sections deal with LNG transportation costs and patterns and likely future developments in LNG tankers.

## INTRODUCTION

The carriage of liquefied natural gas (LNG) at sea dates from 1959, when the 5000 cu.m *Methane Pioneer*, a converted dry cargo ship, operated successfully between Lake Charles, Louisiana, and Canvey Island. This prototype operation led to the first commercial traffic from Arzew, Algeria, to Canvey Island, and to Le Havre, France in 1964 and 1965 respectively, both of which are still in regular operation and are giving every indication of continuing successfully even beyond their 15-year contractual lives.

The growth in the world LNG tanker fleet, firm and estimated, is shown in Fig. 1. As is common with the adoption of any new technology, the beginnings were slow: five and a half years elapsed between the *Methane Pioneer's* historic voyages and the commencement of the first commercial traffic and then there was a further five year gap before the next burst of activity, with the entry into service of *Polar Alaska* and *Arctic Tokyo* in 1969-70 and the delivery of the *Esso Brega, Esso Portovenere, Esso Liguria, Laieta, Descartes,* and *Hassi R'Mel* in 1970-71.

From that point in time the growth was dramatic and from a study of the

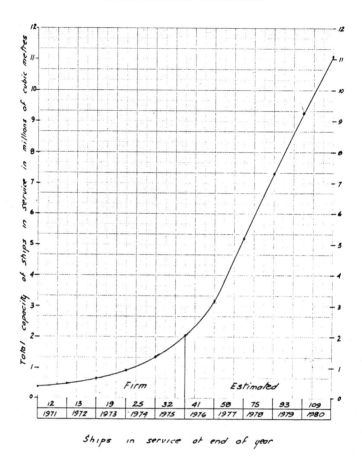

Fig. 1 World LNG tanker fleet, firm and estimated 1971-80

LNG tankers currently on order it appears that by the end of 1975 there will be 32 LNG tankers in service having a total carrying capacity of over 2,000,000 cu.m.

The estimated future growth shown in Fig. 1 is based on an assessment of LNG tankers which will have to be commissioned up to the end of 1980 to satisfy LNG traffic which, in the authors' opinion, are likely to be implemented. A summary of present and projected LNG operations is set out in Table I. The only factors which are likely to prove these estimates over-optimistic are adverse political factors and the discovery of new major gas fields in the very near future in or adjacent to the U.S.A. or Japan, these being the major LNG markets for this decade.

## TABLE I
### Summary of Present and Projected LNG Operations

| From | To | MMCFD | No. Ships | M3 size | Operational | Producer | Liquefier | Shipper | Regasifier |
|---|---|---|---|---|---|---|---|---|---|
| **OPERATIONAL LNG PROJECTS** | | | | | | | | | |
| Arzew | Canvey | 100 | 2 | 27,500 | 1964 | Sonatrach | Camel | British Methane (50% Conch) 50% Gas Council | Gas Council |
| " | Le Havre | 50 | 1 | 24,800 | 1965 | " | (60% Sonatrach) 40% Conch | Gaz de France | Gaz de France |
| Kenai | Tokyo | 140 | 2 | 71,500 | 1969 | Phillips/Marathon | P/M | P/M | Tokyo Gas / Tokyo Electric |
| Marsa El Brega | La Spezia | 235 | 3 | 40,000 | 1970/72 | Esso | Esso | Esso | Snam |
| " | Barcelona | 110 | 1 | 40,000 | 1970/71 | " | " | " | Gas Natural |
| **LNG PROJECTS FIRMLY COMMITTED AND UNDER CONSTRUCTION** | | | | | | | | | |
| Brunei | Japan | 660 | 7 | 75,000 | 1972 | Shell | 45% Shell 45% Mitsubishi 10% Brunei Govt. | 50% Shell 50% Mitsubishi | Tokyo Gas / Tokyo Electric / Osaka Gas |
| Skikda | FOS | 350 | 2 | 40,000 | 1972 | Sonatrach | Sonatrach | Sonatrach Gaz de France | Gaz de France |
| **LNG PROJECTS FIRMLY COMMITTED BUT AWAITING GOVERNMENT O.K.** | | | | | | | | | |
| Skikda | Everett | 50 | 1 | 50,000 | 1972 | Sonatrach | Sonatrach | Alocean | Distrigas |
| " | Staten Island | 100 | 1 | 120,000 | 1974 | " | " | 50% Gazocean 50% Sonatrach | 33 1/3% gazocean 66 2/3% Cabot |
| Arzew | Cove Point | 650 | 6 | 125,000 | 1975/76 | Sonatrach | Sonatrach | El Paso | Columbia Gas |
| " | Savannah | 350 | 3 | 125,000 | " | " | " | " | Southern Natural Gas |
| **LNG PROJECTS STUDIED IN DEPTH AND UNDER NEGOTIATION** | | | | | | | | | |
| Abu Dhabi | Japan | 300 | 4 | 120,000 | 1976/80 | BP/CFP | BP/CFP | BP/CFP (Bridgestone/Gazocean) | ? |
| Nigeria | U.S. East Coast | 650 | 8 | 120,000 | 1976/80 | Shell/BP | Shell/BP/Govt. | Shell/BP | ? |
| Iran (Kharg) | Japan | 500 | 7 | 130,000 | 1976/80 | NIGC/IOP | NIGC/ERAP/C.ITOH/Phillips | NIGC/ERAP/C.ITOH/Phillips | ? |
| Algeria | Barcelona | 150 | 1 | 35,000* | 1974 | Sonatrach | Sonatrach | Naproli | Gas natural |

## TABLE I (CONTINUED)

| From | To | MMCFD | No. Ships | M3 size | Operational | Producer | Liquefier | Shipper | Regasifier |
|---|---|---|---|---|---|---|---|---|---|
| Iran(Queshm) | Japan | 500 | 7 | 120,000 | 1978/80 | Fuji Oil | NIGC/Marubeni | ? | ? |
| Trinidad | U.S. East Coast | 400 | 3 | 80,000 | 1976/80 | Amoco | Amoco/Peoples Gas/Govt. | ? | Peoples Gas |
| Alaska | W.Coast U.S. | 200/500 | 1/3 | 90/130,000 | 1976/80 | Union? | ? | Pacific Lighting | Pacific Lighting |
| Ecuador | W.Coast U.S. | 500(?) | 3/4 | 120/130,000 | 1976/80 | Ada Oil | Ada/Phillips | ? | Pacific Lighting |
| Venezuela (E) | U.S.East Coast | 500 | 2/3 | 90/120,000 | 1976/80 | CVP | ? | ? | PGW? |
| Venezuela (W) | U.S.East Coast | 500 | 2/3 | 90/120,000 | 1976/80 | Esso/CVP? | Esso/CVP ? | Esso ? | Columbia Gas |
| Sarawak | Japan | 700 | 8/6 | 75/100,000 | 1976/80 | Shell | Shell/Mitsubishi/Govt. | Shell/Mitsubishi | ? |
| **LNG PROJECTS STUDIED** | | | | | | | | | |
| N.W.Australia | Japan | 500 | 4 | 120,000 | 1976/80 | Burmah Group† | Burmah Group | Burmah Group? | |
| ,, ,, | ,, | 500 | 4 | 120,000 | 1976/80 | ,, | ,, | ,, | |
| Algeria | N.W. Europe | 500 | 2 | 130,000 | 1976/80 | Sonatrach | ? | ? | ? |
| **LNG PROJECTS LIKELY TO BE DEVELOPED BY 1980** | | | | | | | | | |
| N. Africa | U.S.East Coast | 1000 | 9 | 125,000 | 1976/80 | Sonatrach | ? | ? | ? |
| ,, | ,, | 1000 | 9 | 125,000 | 1976/80 | ,, | ? | ? | ? |
| Skikda | Bilbao | 100 | 1 | 29,000* | 1975/9 | ,, | Sonatrach | Naproli | Petrogas |

* These two ships may become 1 x 75,000 m3 vessel

† Woodside 33⅓%    Burmah 16⅔%    BP 16⅔%    Shell 16⅔%    Chevron 16⅔%    (Burmah owns 31% of Woodside)

In the light of this history there is today little doubt on the technical feasibility and safety of LNG transportation at sea and in this connection it is appropriate here to quote from the "Initial Brief of Commission Staff", dated 16 August 1971, in the Columbia LNG Corporation and others' application before the Federal Power Commission of the U.S.A.[1] This application deals with the import of 1 billion MBtu/day of LNG from Algeria to the U.S. East Cost over a period of 25 years. The Commission staff stated:

"The evidence of record supports the feasibility of delivering the LNG volumes proposed with fleet operations as contemplated by El Paso. While the 125,000-cu.m tankers to be built by El Paso are larger than any currently in service, under construction, or on order, the requirement of Coast Guard approval tends to assure adequacy of design. The feasibility of ocean transportation of LNG has been proven by shipping operations between Sonatrach's (sic) CAMEL facilities, British Methane, and Gaz de France, and in the Phillips–Marathon trade between Alaska and Japan".

## LNG TANKER DESIGN

The carriage of natural gas at sea is effected with the cargo liquefied and at a little above atmospheric pressure (16 psia). The cargo's temperature under these conditions is about $-160^{\circ}C$, depending on the precise composition of the liquefied natural gas. There have been proposals and prototype tests to carry natural gas at sea in other physical states but none have been found to be economically feasible to date.

The main requirements of any design of LNG tanker carrying a cargo at about $-160^{\circ}C$ as a boiling liquid at near atmospheric pressure are:

1. To protect the steel structure of the tanker from the low temperature, and
2. To minimize loss of gas by heat ingress.

The various designs available to achieve these objectives are summarized in Table II, which also identifies the vessels in operation and on order to each of the designs. The most definitive and up-to-date publication reviewing the details of these designs is "LNG Carriers: The Current State of the Art", by Thomas and Schwendtner[2] and it is not intended to go over the ground covered there again. However, for the benefit of those who are unacquainted with the latest developments in LNG tanker design, there follows a brief summary of the main characteristics of each of the designs set out in Table II, and since LNG tanker design is already developing its own vocabulary some definitions will not be out of place.

1. A self-supporting tank design is one in which the LNG is contained in a tank which has in itself sufficient structural strength to withstand the loads imposed on it by the cargo. It is sometimes called "a free-standing tank design". There are four available types of self-supporting tank

according to shape: namely, prismatic (single-wall or double-wall), spherical, and cylindrical.

## TABLE II

### Classification of LNG Tanker Designs

| | | |
|---|---|---|
| | - Single-walled prismatic | -*Methane Pioneer, Methane Princess, Methane Progress, and Heriot**  |
| | - Double-walled prismatic | - *Esso Brega, Esso Portovenere, Esso Liguria, and Laieta* |
| Self-supporting tanks | - Spherical | - *Euclides*, Hull Nos. 176, 177, 196, 197 and 198 (Moss Rosenberg)* |
| | - Cylindrical | - *Jules Verne* |
| | - Single metallic membrane | - *Pythagore, Descartes, Charles Tellier*, Gadania*, Gadila*, Gari*, Gastrana*, Gouldia*, Benjamin Franklin*,* Hull No. 302 (La Ciotat)* |
| Membrane tanks | | |
| | - Double metallic membrane | - *Polar Alaska, Arctic Tokyo, Hassi R'Mel, Geomitra*, Genota*,* Hull Nos. 283, 284 and 287 (France Gironde)* Hull Nos. 1401 and 1402 (La Seyne) |
| Semi-membrane tanks | | *Ethylene Dayspring, Ethylene Daystar* |

* Under construction or on firm order

2. A membrane tank design is one in which the LNG is contained in a thin metallic liquid-tight lining which is supported completely by a load-bearing insulation which in turn is supported by the ship's structure. It is sometimes called "an integrated tank design". There are two types of membrane tank design, one in which there is only a single metallic lining and one in which there are two metallic linings separated by insulation.

3. A semi-membrane tank design differs from a membrane design only in that the metallic lining is not completely supported by the load-bearing insulation but is free of support at the corners. It is sometimes called "a semi-integrated tank design".

All of these designs require a double-hulled ship, the space between the two hulls being used for ballast and the cargo tanks being sited within the holds formed by the inner hull and transverse bulkheads.

### Single-Walled Prismatic Self-Supporting Tank Design

A typical single-walled prismatic self-supporting tank design is illustrated in Figs. 2(a), 2(b), and 2 (c).

To date the cargo tank for this design has been made of aluminium alloy (5083-0), although 9 per cent nickel steel could be used. General particulars

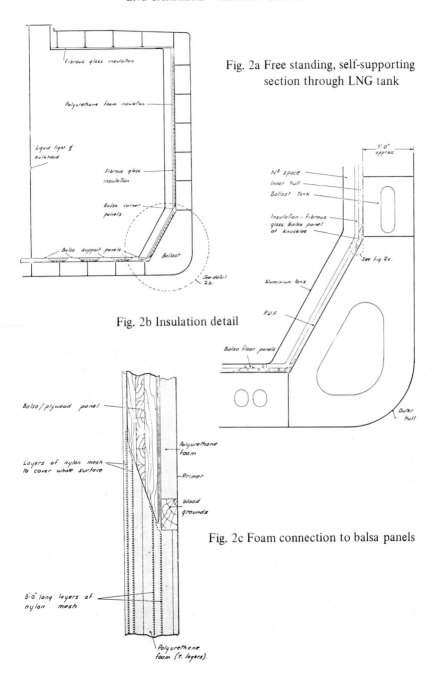

Fibrous glass insulation

Polyurethane foam insulation

Liquid tight ℄ bulkhead

Fibrous glass insulation

Balsa corner panels

Balsa support panels

Ballast

See detail 2b.

**Fig. 2a** Free standing, self-supporting section through LNG tank

N² space
Inner hull
Ballast tank

Insulation: fibrous glass. Balsa panel at knuckles

7'·0" approx.

See Fig 2c.

Aluminium tank

P.U.F.

Balsa floor panels

**Fig. 2b** Insulation detail

Outer hull

Balsa/plywood panel

Layers of nylon mesh to cover whole surface

Polyurethane foam

Primer

Wood grounds

**Fig. 2c** Foam connection to balsa panels

3'·0" long layers of nylon mesh

Polyurethane foam (7. layers).

of typical LNG tankers of varying capacities and employing this design are
given in Table III.

TABLE III
Typical LNG Tanker Single-Wall Prismatic Tank Design

| Cargo capacity (cu m) | | | 90,000 | 105,000 | 120,000 | 130,000 |
|---|---|---|---|---|---|---|
| Length of tanker between perpendiculars (ft) | | | 800 | 840 | 878.5 | 900 |
| Beam | " | " | 124 | 130 | 136 | 139 |
| Depth | " | " | 87.25 | 91 | 95.3 | 97.3 |
| Draft | " | " | 33.0 | 34.5 | 35.75 | 36.5 |
| Number of aluminium alloy cargo tanks | | | 5 | 5 | 6 | 6 |
| Total weight of aluminium alloy cargo tanks (long tons) | | | 3.300 | 3,850 | 4,400 | 4.750 |
| Service speed (knots) | | | 20 | 20 | 20 | 20 |
| Shaft horsepower (service) | | | 36,000 | 39,000 | 41,100 | 42,300 |
| Approximate size of equivalent crude oil tanker (deadweight tons) | | | 83,500 | 96,000 | 110,000 | 119,000 |

The insulation sytem, which is itself the secondary barrier, consists of
balsa/plywood panels along the corners and for the island tank supports
with a foamed-in-place high density polyurethane covering the inner hull
between the corners and the island panels. Fibreglass covers this balsa/
polyurethane foam system.

The tank itself has vertical and horizontal reinforcement members on
the inner surfaces of the tank walls and has a liquid-tight bulkhead fore and
aft and a transverse swash bulkhead. These bulkheads are there for structural
strength but also serve to reduce the forces generated by the free-surface
effect of the cargo at sea and are not possible in membrane designs, where
the free surface effect has to be dealt with in other ways. The tanks are
located by keys and keyways at the bottom and top.

**Double-Walled Prismatic Self-Supporting Tank Design .**

A typical double-walled prismatic self-supporting tank design is illustrated
in Figs. 3 (a) and 3(b). In this design the outer wall of the tank is itself the
secondary barrier and the insulation, which consists of polyurethane or
PVC foam, is attached to the outer wall. The two walls of the tank are
separated and stiffened by T-shaped extrusions. The tanks are fabricated of
aluminium alloy 5083-0 and are located by keys on the centre lines of the
side and end tank walls. These tanks are also sub-divided by internal bulkheads, as in the case of the single-walled prismatic design.

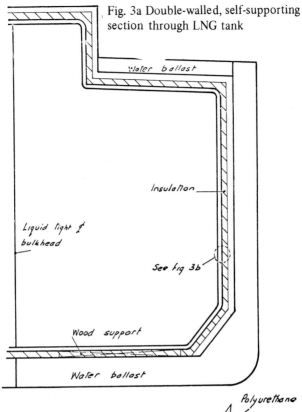

Fig. 3a Double-walled, self-supporting section through LNG tank

Water ballast

Insulation

Liquid tight ¢ bulkhead

See fig 3b

Wood support

Water ballast

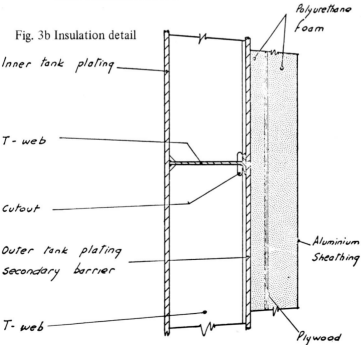

Fig. 3b Insulation detail

Polyurethane foam

Inner tank plating

T-web

Cutout

Outer tank plating
Secondary barrier

T-web

Aluminium Sheathing

Plywood

## Spherical Self-Supporting Tank Design

A typical spherical self-supporting tank design is illustrated in Fig. 4. The tanks consist of a sphere of 9 per cent nickel steel or aluminium and general particulars of typical LNG tankers employing this design are given in Table IV.

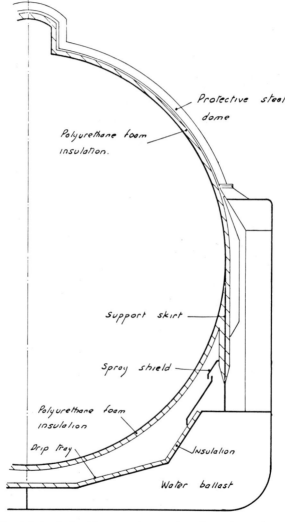

Fig. 4 Spherical self-supporting section through LNG tank

Two systems are being used for supporting the spheres within the inner hull of the vessel. In the one shown in Fig. 4 the sphere is attached to a cylindrical skirt at the equator and the skirt is attached to the ship's structure. The second method of support, employed in the *Euclides*, comprises a system of articulated rods and arms attached at the top to the under-deck structure of the vessel and at the bottom at the equator of the sphere.

## TABLE IV
### LNG Tankers: Spherical Self-Supporting Tank Design

| | | |
|---|---|---|
| Cargo capacity (cu m) | 87,600 | 125,000 |
| Length of tanker between perpendiculars (ft) | 777 | 925 |
| Beam  "        "                          " | 131 | 136 |
| Depth "        "                          " | 75 | 82 |
| Draft  "        "                          " | 34 | 36 |
| Number of spherical tanks | 5 (9% Ni steel) | 6 (5083-0 aluminium) |
| Service speed (knots) | 19.5 | 19 |
| Shaft horsepower | 30,000 | 40,000 |

In the design shown in Fig. 4 the main insulation consists of poly-urethane foam covering the whole outer surface of the sphere and the upper portions of the skirt to control temperature gradients at the upper, critical part of the skirt. An insulated drip-tray system below the sphere is used to provide protection against small leaks.

### Cylindrical Self-Supporting Tanks
The only example of this design constructed to date is the *Jules Verne*, a typical section of which is shown in Fig. 5.

The vertically arranged cylindrical tanks have an inverted truncated cone at the bottom and internal horizontal reinforcing rings. The tanks are fabricated of 9 per cent nickel steel and are supported at the bottom on load-bearing PVC foam insulation covered by a 9 per cent nickel steel tray which acts as a secondary barrier, with a further layer of PVC foam separating the secondary barrier from the primary tank. The side and top insulation consists of PVC foam and perlite, with the secondary barrier at the sides comprising a complex of sheet aluminium, polyethylene film, and cotton cloth between the PVC and the perlite.

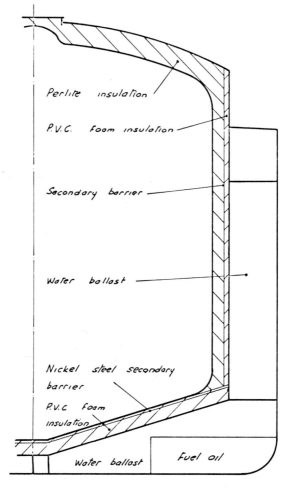

Fig. 5 Cylindrical self-supporting section through LNG tank

## Single Metallic Membrane Tanks

Details of this design are illustrated in Figs. 6(a) and 6(b).

The primary barrier consists of corrugated sheets 1.2 mm thick, made of stainless steel with a very low carbon content 18 per cent chromium and 10 per cent nickel. To allow for contraction in the cold condition, the sheets have two sets of orthogonal corrugations of ogival shape which cross each other by means of special geometrical surfaces which are called "knots".

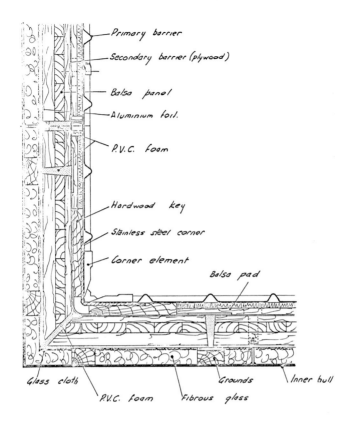

Fig. 6a Membrane design, corner insulation detail

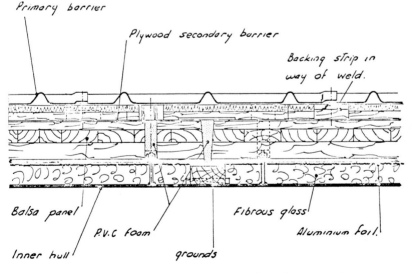

Fig. 6b Membrane design, side wall insulation detail

The sheets are fixed to the supporting insulation along half their perimeter by welding on to small steel blocks embedded in the insulation structure. Furthermore, the sheets are welded together, each sheet being in relation to the adjacent sheet a covering sheet for one of its half perimeters and a covered sheet for the other. The insulation system which incorporates the secondary barrier is composed of plywood and balsa panels, as shown in Fig. 6(b). The plywood and balsa panels are joined together by expanded PVC compressed and bonded to the side edges of the panels and covered by a plywood scab.

Because such tanks cannot support internal bulkheads, the shape of the tank in its upper area departs from a general prismatic shape into a long chamfer which serves to reduce the free surface effect when the tank is substantially full. It is a feature of membrane tanks that in operation they should be either substantially full or substantially empty.

### Double Metallic Membrane Tanks

Details of this design are illustrated in Figs. 7(a) and 7(b). Both the primary and secondary barriers in this case are constructed from Invar (36 per cent nickel iron alloy), which has a very low coefficient of expansion. The sheets from which these membrane barriers are constructed are 0.5 mm thick and are welded along upturned edges horizontally around the periphery of the tank. Otherwise the membranes are flat. The insulation, which is sited between the inner hull and the secondary barrier and between the secondary barrier and the primary barrier, is made of plywood boxes filled with perlite.

As with single metallic membrane tanks, these tanks cannot support interior bulkheads and they are therefore shaped with a long chamfer in their upper region, as shown in Fig. 7(a).

### Semi-Membrane Tanks

Tanks of this type have not yet been used for LNG tankers but they clearly have a potential in such use. They have been applied to small ethylene tankers and the design is now on offer for LNG tankers. The system involves the use of a relatively heavy membrane 4-8 mm thick and few details have been published of the design.

## AVAILABILITY AND COSTS

It will be appreciated that each of the above designs has involved considerable development costs and, as a result, each of the firms who carried out the research, testing, and prototype development are attempting to recover these costs by licensing the techniques involved, all of which are covered by patents in most industrial countries. Generally, the licences are granted to shipbuilders but in a few cases have been granted to oil and gas companies who intend to have constructed sizeable LNG tanker fleets.

Although the first LNG tanker design and prototype originated in the

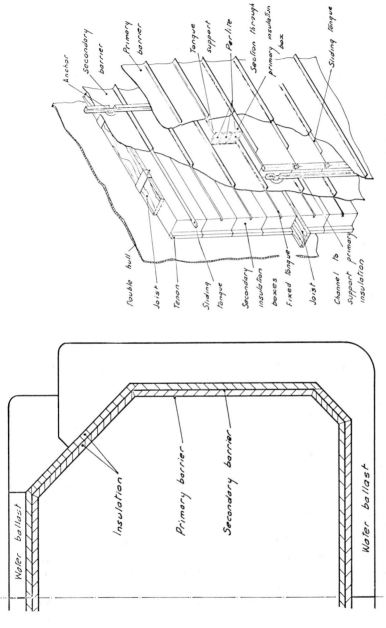

Fig. 7b Insulation detail

Fig. 7a Double membrane design, section through LNG tank

U.S.A. in the late 1950s, European shipyards have built or are building all of the commercial-sized LNG tankers and in Europe it is France that has obtained the vast bulk of the orders. Table V shows details of LNG tankers constructed or on order, broken down geographically. Since orders which are not contractually binding are frequently reported in the press, the data in Table V includes only those orders where a hull number has been allocated.

### TABLE V
### LNG Tanker Construction by Country

| Country | Number of LNG tankers constructed or on order | Carrying capacity (cu m of total LNG tankers constructed or on order |
|---|---|---|
| France | 20 | 1,355,100 |
| Italy | 3 | 120,000 |
| Norway | 5 | 358,200 |
| Spain | 1 | 40,000 |
| Sweden | 2 | 143,000 |
| United Kingdom | 2 | 54,800 |
| U.S.A. | 1 | 5,100 |
| World | 34 | 2,076,200 |

In the last year there are signs that the domination of French shipbuilders, in particular, in LNG tanker construction may be eroded by serious competition from the U.S.A. and Japan. In the U.S.A. there is growing political pressure to ensure that any base-load LNG imports are carried in tankers constructed in the U.S.A. and carrying the U.S.A. flag. This political pressure has resulted in substantial sums being allocated to the U.S. Maritime Administration for subsidizing U.S.-built LNG tankers. Thus, for the fiscal year 1973 (July 1972 — June 1973) 72.2 million dollars are available to the Maritime Administration for subsidizing four LNG tankers. This amounts to a subsidy of about 22.5 per cent per tanker and it is interesting to compare this with the 43 per cent subsidy that is currently being granted by the U.S. Maritime Administration on oil tankers, which leads to the conclusion that the differential between U.S. building and non-U.S. building for LNG tankers is substantially less than that for an oil tanker. Whether this, in fact, proves to be true is yet to be seen, because no firm construction contracts for LNG tankers have yet been let in the U.S.A.

As regards Japan, the shipbuilders there were loth to enter the specialized field of LNG tanker construction during the 1960s, when they could fill their order books with convential tankers. The position there changed radically in 1971, when the currency crisis caused a falling-off of tanker orders and all the major Japanese shipbuilders are currently evaluating the various LNG tanker designs and are putting themselves in a position to bid on such tankers from the middle of 1972 onwards.

It is clear, therefore, that we are entering into a phase where there will be considerably greater competition for the construction of LNG tankers and owners will have a much wider choice of yards to go to. It is expected that this will help hold down the rapid increase in the cost of LNG tankers which has been characteristic of the last five years. There is little doubt that the shipyard capacity in Europe, Japan and the U.S.A. is capable of meeting the expected LNG tanker demand in this decade.

This brings one on to the prices of LNG tankers and the factors affecting price. This is a very difficult subject, because shipyards and owners seldom publish this data with any degree of accuracy. From time to time, prices are published in the press, but they are usually meaningless because they do not specify what elements are included and what excluded from the published price or even whether it is a delivered price or a base price with escalation provisions. Price comparisons between the various designs are even more difficult because usually they have to be made from quotations from different yards in different countries with different labour rates, productivity, overheads, profit margins, subsidies, etc., and even if the quotations on the different designs come from a single yard they may not be strictly comparative in a general sense because the yard's facilities may be more suited to the building of one design than another. What can be done here is to examine the factors in the various designs that give rise to price differentials between them and to add some specific reliable prices which have been recently published.

As regards the price element due to the hull and machinery, there is little difference between the various designs. The membrane designs use the cubic capacity of the vessel most effectively and therefore can employ the smallest hull for a given carrying capacity. The prismatic self-supporting tanks use the cubic capacity almost as well as the membranes but in the case of cylindrical or spherical tanks a longer or beamier hull is necessary for a given carrying capacity and therefore the price element attributable to the hull and machinery is higher. Also, cylindrical or spherical tanks involve penetration of the deck by the tanks, which results in the need to further stiffen the deck involving additional steel.

The price element due to the material of the primary LNG containment system is considerably lower in the case of the membrane designs than in the case of the self-supporting tank designs. It will have been noted in Table III that the total weight of the aluminium alloy cargo tanks in a prismatic self-supporting tank design LNG tanker of 120,000 cu.m capacity is something over 4400 long tons. In contrast, the weight of metal in a membrane LNG tanker of this capacity is probably a little less than 300 tons.

These differences in material weight and, therefore, cost, however, are offset to some extent by the possibility of using cheaper insulation systems with the self-supporting tank design than are possible in a membrane design, where the insulation system has to take the full loads imposed by the cargo. Of course, in addition to the material costs of the primary LNG containment system there is the fabrication cost and this is a subject on which little has been published and which varies considerably from country to country and with the technical requirements of the system, *e.g.* the extent to which automatic welding can be used and the method of weld testing involved. Moreover, there may well be substantial differences in price generated by use of selected materials for substantially the same design. Thus, some authors have claimed that the use of 9 per cent nickel steel in the place of aluminium alloy for the single-walled self-supporting tank design can result in a 10 per cent saving on the cost of the tanks[3].

An advantage of the self-supporting tank designs which should affect price favourably is the fact that the tanks can be shop-built at the same time as the hull is under construction, whereas the membrane designs need at least a partially completed hull before the insulation and membrane can be installed. Although this can be done in a fitting-out dock, so that there need be no difference in time in the building dock, the overall period of construction for the membrane designs is almost inevitably somewhat longer than for the self-supporting designs, which generally require two and a half years building time.

The cargo-handling element of the price (pumps, liquid and vapour lines, cooldown systems, instrumentation, etc.) varies little from design to design.

To summarize on this question of price, the consensus of opinion is that self-supporting tank designs are somewhat more expensive than the membrane designs, primarily because they require considerably more special alloy for the containment system. However, by their very nature, they are more rugged and easier to analyse and less likely to suffer any damage to the containment system over the life of the vessel. No case of rupture of the primary barrier has ever been encountered on LNG tankers of the self-supporting type, whereas such ruptures have occurred at least twice on one of the membrane systems. It is therefore necessary to look, not only at the initial capital cost of the vessels, but also at what has been called their "life-cycle cost" (see ref 2).

## LIFE-CYCLE COST

A natural gas liquefaction and transportation project comprises the following operations:
1. Gas production
2. Gas pipeline

3. Liquefaction plant and tanker loading terminal
4. LNG tankers
5. Tanker receiving terminal and vaporization.
Such a complex is very capital-intensive and therefore continuity of
operation is of paramount importance. LNG tanker outages of a few days
duration may be gradually made up by utilizing whatever limited margin
may be built into the various operations of the project, provided that the
time intervals between the stoppages caused by tanker outages are not too
short. On the other hand, if unscheduled tanker outages of extended
duration cannot be made up, there will be a commensurate loss of sales
revenue to all participants in the venture. Almost all project costs will be
virtually unaffected by a tanker outage and therefore, for the venture as a
whole, the loss of sales revenue for vaporized LNG represents a direct loss
of income. This aggregate loss would be distributed among the operating
companies according to the various transfer prices.

For the present purpose let us say that the price of vaporized gas ex
receiving terminal is 1.00 U.S. $/MMBtu. A specific ship operating on a
given project has the job of delivering a nominally fixed quantity of Btu/
year from the LNG plant to the receiving terminal. This annual delivery is
usually assumed to be made in 345 calendar days, including port times,
with allowance for bad weather and other minor delays. We assume that if
the vessel has an extended unscheduled outage of $Z$ days, the loss in sales
revenue to the venture as a whole if not made up is given by:
$\frac{Z}{345}$ x the vessel's assigned annual MMBtu deliverability quantity
          x 1.00 U.S. $.

On this basis the sales revenue loss, as a function of unscheduled ship
outage time, has been worked out for a 120,000 cu.m ship on various journeys
and is graphed in Fig.8. It should be borne in mind that unscheduled ship
outage time may take the form of reduced carrying capacity.

From the data in Fig.8 we can calculate the break-even unscheduled
outage time for an LNG tanker which would justify, say, a 10 per cent
higher capital investment if that led to complete avoidance of such outage
time. This has been done in two ways:
(a)   As a single outage (presumed to occur at the beginning of the ship's
      life and therefore taking no account of the time value of money); and
(b)   As a number of equal annual periods of outage occurring throughout
      the vessel's 20-year contract and which allow for the 10 per cent
      extra initial investment being recovered with an 8 per cent interest
      cost included.

The results are given in Table VI. It is to be noted that the shorter-haul
projects are more sensitive to unscheduled tanker outage time than the
larger ones.

In the foregoing it has been assumed that sales lost by prolonged unscheduled ship outages are not subsequently made up. Of course, this could be

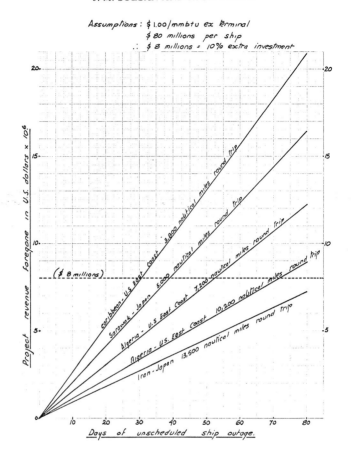

Fig. 8 Project revenue foregone as a result of unscheduled ship outage

catered for by building larger ships, but this involves not only additional
first cost of the LNG tankers but will also involve additional first cost for
LNG storage and/or liquefaction capacity.

Fig.8 and Table VI show how sensitive an LNG project is to unscheduled
outage time in the tankers. Unscheduled outage time due to defects in the
conventional parts of the tankers and in those parts that are common to all
containment systems, such as the pumps, instrumentation, and other cargo
handling features, are, of course, not relevant in selecting between different
containment systems. The major item that is a factor is unscheduled outage
time due to a rupture of the primary barrier. While statistics on this point

## TABLE VI
## Break-even Unscheduled Ship Outage Time for Various LNG Projects

| Project | Round trip, nautical miles | Break-even single unscheduled Ship Outage to recover 10% extra initial ship investment (Days) | Break-even average unscheduled Outage time per contract year to recover 10% extra initial ship investment at 8% interest (Days per year – 20 years) |
|---|---|---|---|
| 1 | 2 | 3 | 4 |
| A. Caribbean - U.S. East Coast | 3,800 | 31 | 3.0 |
| B. Sarawak - Japan | 5,000 | 39 | 3.8 |
| C. Algeria - U.S. East Coast | 7,200 | 53 | 5.2 |
| D. Nigeria - U.S. East Coast | 10,200 | 72 | 7.1 |
| E. Iran - Japan | 13,500 | 88 | 8.7 |

Notes:

1. 120,000 cu.m ships costing $80 million each are assumed for all projects.

2. $1.00/MMBtu ex re-gasification terminal is assumed for each destination.

3. Column 3 represents an undiscounted recovery.

4. Column 4 represents a recovery over a 20-year contract at an 8 per cent deferment rate and assuming regular equal outages.

are not yet available on all containment systems, because some have not yet operated at sea at all and others have only operated at sea for short periods, the following can be said with certainty: there has never been a rupture of the primary barrier in any self-supporting tank system, but one of the membrane has had at least two ruptures. This fact must be reviewed in the light that, considering ships in regular commercial operation, the self-supporting systems have a total of 25 ship-years LNG operation, whereas the membranes have only four ship-years in LNG operation.

## LNG TRANSPORTATION COSTS

One of the most reliable sources of both capital and operating costs of LNG tankers is the Federal Power Commission in the U.S.A. In the Columbia LNG Corpn *et al* application[1], the initial brief of the Commission's staff gives $563,800,000 as the capital cost estimate of a nine-ship fleet of 125,000-cu.m LNG tankers, *i.e.* an average of $62,644,000/tanker. This price included three comparatively inexpensive vessels to be built in France, the construction contracts of which were placed in 1970. Similar vessels contracted for in 1972 in France would probably cost $70 million and in the U.S.A. well over $80 million. As far as can be established, these costs do not include interest during construction or commitment fees and such charges would add between 10 and 15 per cent to the ultimate cost to the owner. As regards operating costs, Table VII gives a typical breakdown of estimated operating costs for the year 1977 for a 125,000-cu.m LNG tanker. This is confirmed again in ref 1, where the annual operating cost for the nine-vessel fleet was given as just under U.S. $32 million at 1971 cost levels, *i.e.* $3,555,000/tanker.

### TABLE VII

### LNG Shipping Charges

| Source | Destination | Distance (nautical miles) | Quantity (MMCFD) | Shipping Price (U.S. cents/ MMBtu) |
|---|---|---|---|---|
| 1. Marsa el Brega | La Spezia | 990 | 235 | 8.5 |
| 2. Marsa el Brega | Barcelona | 1060 | 110 | 9.0 |
| 3. Marsa el Brega | New York | 4540 | 30 | 40.5 |
| 4. Arzew | Cove Point | 3600 | 650 | 33.40 |
| 5. Skikda | New York | 3470 | 150 | 31.50 |
| 6. Iran | Japan | 6770 | 400 | 59.00 |

Shipping charges for LNG operations, being a combination of capital charge and operating costs, have escalated rapidly over the last few years owing to the considerable escalation of both elements. Typical published shipping charges for some existing and projected schemes are set out in Table VIII.

## TABLE VIII

### Operating Costs (1977) for a 125,000-cu.m LNG Tanker

|                                        | Annual cost (U.S. Dollars) |
| -------------------------------------- | -------------------------: |
| Payroll                                |                    426,000 |
| Subsistence                            |                     26,000 |
| Deck, engineering, and steward stores  |                     60,000 |
| Repairs and maintenance                |                    275,000 |
| Insurance (vessel)                     |                  2,014,000 |
| Insurance (cargo)                      |                     30,000 |
| Technical assistance                   |                     25,000 |
| Crew travel and miscellaneous          |                     44,000 |
| Management                             |                     70,000 |
| Port charges                           |                     88,000 |
| Fuel oil                               |                    694,000 |
| Nitrogen                               |                     65,000 |
|                                        |                  3,817,000 |

It is interesting to compare the cost of LNG transportation with that of crude oil for varying distances and this is shown in the curves in Fig.9. It will be seen that the cost of moving energy as LNG is about four times that of moving it as crude oil for any given distance.

## LNG SHIPPING PATTERNS

It is generally accepted that all the phases of an LNG project have to be co-ordinated so that the required quantity of LNG is delivered most economically. This co-ordination is carried out at the initial study phase through an evaluation of alternative combinations of various components of the project which have a direct effect on each other. The components to be considered are the number, size, and speed of ships, the capacity of the LNG storage both at the plant and the receiving terminal, the loading facilities involved, and dredging requirements, if any. From such a study the optimum number, size, and speed of ships is determined.

This is the current pattern of LNG tanker procurement, *i.e.* specifically-sized tankers optimized and built into the other specific features of the whole project, and this pattern in general is likely to remain so. Moreover, these LNG tankers have been owned by corporate entities involved in the LNG scheme as a whole.

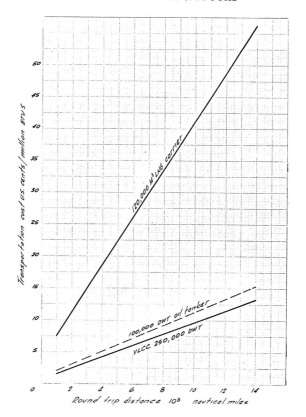

Fig. 9 Comparative costs for energy carried by sea as LNG or as crude oil.
Basis: 1975-76 costs and same financial charge.

However, the pattern outlined above is likely to show some changes in
the near future. LNG tankers are being ordered on speculation and the trade
for them procured during construction. The first example was the *Descartes,*
the 50,000-cu.m LNG tanker ordered in October 1968 and which went into
service in October 1971. Because of the timing of its order, it is probably the
least expensive LNG tanker in terms of capacity currently afloat. Since the
ordering of the *Descartes,* at least seven LNG tankers, varying in size from
29,000-cu.m to 125,000-cu.m capacity, have been ordered on speculation. These
orders are placed by shipping companies as distinct from LNG traders and are
possible largely because, with rapid escalation of costs, the sooner vessels
are ordered the lower the final cost, even if the construction contract contains
escalation clauses on some elements of the price. Moreover, the overall
capital requirements of LNG schemes now being put into operation are so
vast that the LNG traders will be only too happy to avoid part of them by

time-chartering vessels from regular shipping companies. It is believed that there will be a growing tendency for LNG fleets to be partly owned and partly chartered or even wholly chartered for this reason.

The speculative owner is, of course, faced with the problem of the size of vessel which he should build, but this is not too serious because the potential LNG schemes for the next decade are all well known and it is not too difficult to arrive at accurate conclusions as to the size of vessel that each of them will require. The great danger to the speculative owner is misjudgment of timing *vis-a-vis* the availability of liquefaction capacity.

## FUTURE DEVELOPMENTS OF LNG TANKERS

It is not likely that any major breakthrough in technology of LNG tanker designs will occur in the next five years, particularly since over half of the designs currently on offer have not yet been tested at sea. It is therefore unlikely that improvements in design will result in any reduction in capital cost and, in the authors' opinion, is barely likely to keep pace with inflation. In this period, therefore, the only economies that will arise in LNG transportation will be those arising from increasing the size of the individual LNG tankers. At the moment, the largest LNG tankers in operation have a capacity of 71,500-cu.m. Some designers have available specifications for LNG tankers of up to 200,000-cu.m, but it is believed that owners will wish to obtain some experience of the 125,000-cu.m class before making the next step forward. There is no doubt that, particularly on very long runs, substantial economies in transportation costs will be obtained by going to larger tankers but it is unlikely that we will see 200,000-cu.m tankers at sea much before 1980.

As regards completely new designs, a number of companies are working on what is being termed a "wet wall system". This system does away with a metallic containment system for the LNG, which would be contained within layers of a suitable foamed plastic within the inner hull, the foamed plastic acting as barrier and insulation. This technique has already been achieved for LPG carried at a temperature of about $-40^\circ C$ but has yet to be developed for LNG.

## ACKNOWLEDGEMENTS

The authors thank Conch Methane Services Ltd for permission to publish this paper and J. W. Hunt and P. L. L. Vrancken for their valuable contributions to the data included.

## REFERENCES

1. "Initial Brief of Commission Staff". Federal Power Commission. Columbia LNG Corporation *et al.* 16 August 1971, Washington, D. C.
2. Thomas and Schwendtner, "LNG Carriers: The Current State of the Art". Annual Meeting (New York) Society of Naval Architects and Marine Engineers, 11-12 November 1971.

3. Pozzolini, P. F., Scarpa, L., and Raggi, F. "Ottimizzazione del progetto per un serbatoio autoportante per nave metaniera". Convegno di Tecnica Navale, Associazione Italiana di Tecnica Navale, Trieste, 25-27 March, 1971.

## DISCUSSION

*G. A. Harris* (Lambert Brothers Shipping Ltd) said it was a daunting prospect to comment on a paper whose authors claim to have invented part of the terminology of the title. He hoped he could do them justice. Without further prevarication he wanted to congratulate them on their comprehensive survey of an exciting new development which had captured the interest, imagination, and resources of many oil, gas, and shipping companies as well as, of course, the environmentalists.

Before turning to one or two questions, he wished to make two observations. First, perhaps he could be allowed licence to comment further on the Executive Committee's choice of title. The dictionary definition of a liner needed some expansion, as well as regularity of service between two specific ports. There were certain other features of liner operations to which he wished to draw attention. First, a liner carried a variety of commodities; secondly, its customers, the shippers, were many; and thirdly, its freight rates were unilaterally determined. The reason for mentioning these other features, which were demonstrably not applicable to LNG shipping, was not to chide the Committee for a conceptual misunderstanding, but to emphasize the novelty of LNG transportation for the shipping industry.

This brought him to his second observation, which concerned the supply of transportation to meet the forecast demand for LNG carriers and the role of the independent ship owner. It was worth reiterating the principal characteristics of the market, as described by the authors:

1. The shipping demand would be regular throughout the year, rather than seasonal, unlike tanker demand. It would also, in the short term, be predictable.
2. The cost of one LNG carrier was high even by oil company standards. For a vessel of 120,000 cu.m, it was of the order of two to three times the cost of a VLCC.
3. For every LNG project, it was essential to optimize on the number, size, and speed of ships in order to achieve the most profitable operation.

Because of these characteristics, we were unlikely to witness the development of a tramp market, *i.e.* the speculated operation by independent shipowners of vessels on voyage or short-term charters, nor, again, was the speculative ordering of LNG carriers on a large scale an advisable course of action for the independent shipowner (unless the orders were obtained at a comparatively low price), since he ran a considerable risk either of possessing a sub-optimal design of vessel for a given project or of finding the market pre-empted when he attempted to charter out the vessel.

However, it did not follow from these conclusions that the independent shipowner had no part in the action: his operating expertise and capital resources could be of immense value to the LNG shipper. But in order obtain maximum benefit from the participation of the shipowner it was important to establish, in the planning stage of the project, a closer co-operation between owners and charterers than had been normal in the tanker market. In achieving this co-operation, a clear lead and greater dissemination of information must come from the LNG shipper: this had certainly been the consensus of many major shipowners in Western Europe during the course of his recent meetings with them.

He then put some specific questions to the authors of the paper.

1. Was the capital cost likely to fall, or at least stabilize, in the future as a result of (*a*) series production (the opportunity for economy may vary between designs) or (*b*) the wet wall system?

2. Was the design too complex for construction in emerging shipbuilding countries (*e.g.* S. Korea, Brazil) for the foreseeable future?

3. Why had a temporary plateau been reached at 120,000–130,000 cu. m? Were further transportation economies by increasing the size of ship sufficient to outweigh diseconomies of additional plant and storage?

4. Two ruptures of the membrane design of tank were mentioned. Apart from the cost of unscheduled outage time, how high was the cost of repairs on this design, given the need to virtually rebuild the tank? How did that compare with the rupture of a free-standing tank?

5. Did the rate of gas boil-off, which was generally used as a fuel for a vessel, vary between designs? Did this affect the economics significantly? Could the rate be varied on any of the designs to take advantage of any change in the relative prices of fuel oil and natural gas?

*R. C. Ffooks*, in answer to the first question on capital cost, replied that in a "ten-off" situation the capital cost was likely to fall but it was uncertain that a further reduction would be obtained in a "20–50-off" situation. Where the design allowed one to sub-contract some of the work, a 30 per cent reduction in capital cost might be obtained in, say, a "12-off" situation.

At that stage it was difficult to know whether the wet wall system would allow a reduction in capital cost. They did not know what complications may arise.

They were at present seeing prices quoted that were higher than needed, in that shipyards were being asked to carry the risk of these ships, *i.e.* to supply a guarantee to the owners. A large margin was included now because the ships were new and this provided risk cover.

As to the role of the emerging shipbuilding countries in LNG liner construction, in general the current technology did not exist in these countries. However, the technology usually had to be bought. Nevertheless, a large part of the tank work was sub-contracted and they could do this.

As to the current plateau at 120,000–130,000 cu. m, there was technically no reason why we should not have larger vessels. 165,000–

200,000 cu. m designs were now available. However, until they looked at the performance of the new 120,000-cu. m vessels it was unlikely that larger ones would be built. The larger vessels would be more economic, particularly on the longer runs such as to Japan.

Most of the ruptures so far had been local and in this case the cost of repairing the membrane was not particularly high. The real problem was in gas freeing and staging before actually starting on repairs. In a recent case it took four to five days to repair the membrane, but for the total operation the vessel was out for 40 days. The self-supporting type was easier to gas-free. If they were talking of collisions, then this was quite different.

The gas boil-off did vary between designs. The membrane type was more expensive and thermally less effective. It had a higher boil-off than the self-supporting type by almost a factor of two. The self-supporting type could deliver slightly more cargo and might make $100,000—$200,000/year difference in income.

*I. W. Robertson* (H. Clarkson & Co. Ltd) commented on the question of whether prices were likely to come down. They had not gone up dramatically. There was at present about a $25 million spread in asking price between yards with no experience or orders and those who had. There was probably a $10—$15 million margin per ship in builders' prices. This was a built-in safety margin.

On the question of insurance being 53 per cent of running cost, they had calculated some at over 60 per cent. Was this very high insurance figure related to underwriters, considering this was a very dangerous ship compared with a VLCC?

*R. C. Ffooks* said that though the insurance was a large percentage of operating cost, the rate as a per cent of capital cost was exactly the same as for a tanker. He believed the rate was based on this. They would say that these vessels were safer than tankers.

*J. M. C. Bishop* (Phillips Petroleum Co. Ltd) said that in view of sensitivity of lost time on short voyages, would it not be possible to increase the speed and so improve the economics?

*R. C. Ffooks* replied that the operating speed of these vessels was higher than that of tankers. An equivalent tanker had a speed of 16—17 knots, compared with 19½ knots for these vessels.

One had to do the optimization (*i.e.* the speed has been decided) first and then look at reliability.

*B. Davey* (Air Products Ltd) said that if they did not wish to build in reserve speed but have had "outage" time, would they not want to reduce turn-around by increasing pumps.

*J. M. Soesan* assumed they had already optimized but were still subject to outage time.

*J. L. L. Orbach* (Northern and Central Gas Corpn Ltd) asked whether there were any designers of a system who offered effective guarantees, other than Chicago Iron.

*J. M. Soesan* said that the normal guarantee was one year from delivery. One was unlikely to get any yard to offer a guarantee for a longer period.

*G. A. Hogg* (British Petroleum Co. Ltd) said that over the question of speed of LNG carriers, they had gathered that in the Gachsaran/Bibi Hakime/Kharg Island project mentioned in the paper, the National Iranian Gas Company had proposed to employ five 25-knot LNG carriers instead of seven conventional 19-knot vessels. Could Mr Ffooks comment on the practical and economic plausibility of this proposal?

*R. C. Ffooks* replied that as ships got bigger, the most economic would be the bigger ones. A 200,000- cu. m vessel would do over 20 knots, but getting above this speed involved twin-screw problems. This had some bearing on the present size plateau. They were now on the verge of employing twin screws which would involve higher capital and operating costs.

# The Value of Competing Energy Forms and Their Relationship to LNG

By I. A. I. GRIFFITHS and D. G. M. BOYD

*(Shell International Gas Ltd)*

---

Natural gas is liquefied for two main reasons; first, because it may be easier and more economical to store in liquefied form than as gas, and secondly, because it can only be economically transported as a liquid to supply certain demands for natural gas. Since the process of liquefaction, storage, and re-gasification is very costly, as is the transportation of LNG, it is only attractive to producer, transporter, distributor, and end user if it provides gas which will meet a market demand for a high-quality fuel at a price competitive with other high-quality fuels.

At present the commercial situations in which liquefaction is regarded as economic largely centre on the storage of LNG by gas transmission and distribution companies for peak-shaving, and the transportation of natural gas by sea. There are, of course, other situations in which liquefaction provides economic advantages at present or could possibly be seen to do so in the future as, for instance, for fuelling of some commercial vehicle fleets. Further advantages can be gained from LNG by using the cold which it contains and thus reducing the net cost of re-gasification.

It is therefore the purpose of this paper to examine the opportunities for the sale of LNG in the major world markets and the competition from other energy forms which will help to determine the direction of the development of these markets. The value of LNG will be compared to that of other competitive energy forms.

At the time when this paper was being prepared the full effects of the currency re-alignments on the costs of the competing sources of energy was still unclear. Costs are therefore quoted to fall within a wide range. References are always to U.S. cents.

## BASE LOAD SUPPLIES OF LNG

The supply on a regular basis of LNG to the major energy-hungry markets of the world from countries with little present demand for very large volumes of clean energy has rapidly become an important element in plans for meeting future energy deficits. The value of LNG related to competitive sources of clean energy, however, varies considerably in the three major importing areas, U.S.A., Western Europe, and Japan, and it is therefore necessary to deal with the market situation separately in each area.

### U.S.A. Market

In the U.S.A. the role of LNG will be to provide additional pipeline quality gas to meet the enormous deficit in supply of indigenous natural gas against forecast demand. Current forecasts of reserves-to-consumption ratios for indigenous natural gas show a declining trend against an increasing demand for gas both as a domestic fuel for winter heating and for industrial use.

One forecast of indigenous natural gas supply shortfalls made late in 1971 by the National Petroleum Council shows the problem facing energy planners in the U.S.A:

|  | $(10^{12} \text{ ft}^3)$ | | |
|---|---|---|---|
|  | 1975 | 1980 | 1985 |
| Potential gas demand* | 29.32 | 33.61 | 38.87 |
| Indigenous gas supply | 21.72 | 21.77 | 21.49 |
| Resultant gas shortfall | 7.60 | 11.84 | 17.38 |

Although these shortfalls could be substantially lower if reliance is placed on different assumptions, some of which differ radically from those used above, nevertheless all market forecasters agree that by the end of the 1970s and onwards they are likely to be very large.

Relative to total demand, only limited quantities of Canadian natural gas are currently thought likely to be available. The other most porbable sources of supplies of high-quality supplemental gas are LNG and substitute natural gas (SNG) from coal, crude oil, naphtha, or NGL. Apart from imports of gas by pipeline, these sources will be examined in turn.

In the long term, coal gasification may be considered the most attractive solution to the shortage of gas, since dependence upon foreign imports is reduced.

Presently-known mineable reserves in the U.S.A. could supply some $11,000 \times 10^{12} \text{ ft}^3$ of gas. When compared with deficits of even $17 \times 10^{12}$ $\text{ft}^3/\text{year}$, the importance of encouraging this source of gas for the U.S. economy becomes obvious.

* Based on gas retaining its current share of total energy demand.

Several processes are being examined. One advanced process by the Institute of Gas Technology, known as the HYGAS process, has reached pilot stage. Methane is produced by reacting coal directly with hydrogen at high temperatures and pressure to produce two-thirds of the final available gas while one-third is produced by catalytic reaction of hydrogen and carbon oxides. Even if a reasonable time is allowed for the final development of the HYGAS or other competitive processes, it is not likely that a commercial plant will come onstream before 1980, though a crash programme could advance the start-up date by two or three years.

Estimates of the ex-plant price for this gas by 1980, taking cost inflation into account, are of the order of 110-130 cents/M $ft^3$ based on the use of Illinois coal in a plant capable of producing up to $250 \times 10^6$ $ft^3$/day. The effect of further increases of scale is thought to be small, of the order of 2-3 cents/ M $ft^3$ for a $500 \times 10^6$ $ft^3$/day plant. Cheaper Wyoming coal at \$3.30/ ton could reduce the price to around 90 cents/M $ft^3$.

The most significant factor affecting the cost of gas by this route is the cost of coal, which can account for almost half the total cost.

The Federal Government is financially supporting all pilot-plant work and there is a growing feeling that the U.S.A. should exploit its large domestic coal reserves. In 1971 a bill was introduced in the Senate to allow rapid amortization over a five-year period of facilities to convert coal and oil shale into synthetic fuels.

As another benefit of coal gasification additional employment is provided. It has been estimated that over 40,000 new jobs are created by investment in facilities to produce 1 $\times 10^{12}$ $ft^3$ annually.

In the short term substitute natural gas (SNG) from naphtha is attractive because the technology is well known, the time to build the plant is short (about 1½–2 years), and the capital cost of plant is relatively low compared with coal gasification processes. Capital and operating costs are in the region of 15-20 cents/M $ft^3$, giving an ex-plant gas cost of 95-100 cents/M $ft^3$ if a naphtha price of 8½ cents/USG is taken. At present, insufficient supplies of domestic U.S. naphtha are available for the gas companies' needs, but the administration is not as yet prepared to permit imports of foreign naphtha. At late 1971 prices gas from foreign naphtha would cost at least \$1.00/$10^6$ Btu, but the price of foreign naphtha is likely to rise to the point where further rises are constrained by the cost of gas production from imported crude or domestic naphthas. Using fairly conservative assumptions on crude oil prices, the cost of gas from these latter sources would be around \$1.40 to \$1.50/$10^6$ Btu, giving a ceiling for foreign naptha of around 12-14 cents/USG.

Important to any gas utility thinking of buying naphtha on a long-term basis for SNG manufacture will be the certainty of long-term supply. But with demand from the petrochemical industry and the use of naphtha for burning as a low sulphur fuel developing, competition for naphtha could restrict the amount available for gas making.

Several plants have been planned for the manufacture of SNG from natural gas liquids. This process should be trouble-free and SNG costs by this route may not be higher than from naphtha unless the feedstock requires pre-distillation or Gas Recycle Hydrogenation.

SNG manufacture from crude oil or heavy fuel oil is still at a very early stage of development. No large-scale plants have been built as yet, and therefore the costs of SNG manufacture for base load supply remain in doubt.

The two most promising ways of manufacturing SNG from heavy fuel oil and crude appear to be:

Using partial oxidation such as the Shell Gasification.

Process in connection with Gas Recycle Hydrogenation.

The Fluidized Bed Hydrogenation Process.

The former route shows a higher capital cost for the plant to gasify fuel oil and overall gives a higher ex-plant gas cost. Estimates of capital and operating costs (excluding feedstock) for the former process are of the order of 35-45 cents/$10^6$ Btu and, for the latter, 15-25 cents/$10^6$ Btu.

The technology of Fluidized Bed Hydrogenation still remains largely untested, and a prototype plant, belonging to Osaka Gas, only operates intermittently.

Compared with all forms of SNG manufacture mentioned, in which feedstock prices are either the major or a very large component of the gas cost, the cost of gas from LNG is largely affected by the capital cost of the very expensive liquefaction plants, ships, and terminals necessary for its transport. For this reason, delays by the Federal Power Commission in permitting imports can sensibly increase the average delivered cost over a 20-year period at a time of rapidly increasing costs for labour and raw materials. Estimates of prices for LNG coming into the U.S.A., East and West Coasts from Algeria, Libya, Trinidad, Venezuela or Nigeria, Alaska, Canada, Ecuador vary between 77-120 cents/$10^6$ Btu ex-re-gasification plant. The into-liquefaction plant gas cost is of the order of only 10-20 per cent of the total ex-re-gasification cost.

By 1980 the U.S.A. could be importing some $2 \times 10^{12}$ ft$^3$/year of LNG from the sources mentioned, if present plans materialize and the necessary governmental approvals are forthcoming. First large-scale imports are likely to start by 1976-77. Since SNG from coal is not likely to be available in significant quantities before the 1980s, SNG from naphtha is the only other practical source of supplemental gas in the 1970s. There are suggestions that for the present U.S. Administration the most important drawback of imported LNG is in its lack of flexibility of supply, unlike naphtha and crude oil. But in balancing risks against benefits, the Administration will undoubtedly take into account that all LNG schemes are unlikely to be interrupted simultaneously. Moreover, all planned LNG imports are not likely to supply more than a very small percent of total natural gas demand. In this context it is interesting to note that a recent government agency report has indicated that LNG imports will be required.

In the long term the incremental supply of additional gas would seem to be shared between all sources discussed above, with up to $1.5 \times 10^{12}$ $ft^3$ of gas from naphtha and NGL being made available by the second half of this decade, with LNG being available from 1976-77 and with coal gasification in the long term making available the bulk of additional requirements.

Since the costs of LNG are largely determined by the very large initial capital expenditures resulting from the need to liquefy and keep liquid natural gas at cryogenic temperatures, investigations have been made into the feasibility of converting natural gas to crude methanol at source and shipping the resulting product to the market for re-gasification or for use as a fuel with excellent non-pollutant characteristics. The major element of expense is the methanol plant which on the scale necessary involves further development of existing technology. At present, the cost of producing pipeline quality gas by this route looks high by comparison with LNG, but it is too early for this route to be excluded in the long term.

**Japanese Market**

In Japan there is no national gas grid nor countrywide market for piped natural gas. Indigenous production is small (about $75 \times 10^9$ $ft^3$/year) or $<$ 1 per cent of total energy demand. The very large gas companies in Tokyo, Osaka, and Nagoya account for around 80 per cent of gas sold, which is largely for heating and cooking. These companies supply town gas manufactured from many feedstocks. Owing to the high cost of manufactured gas which is twice the price of kerosine, the domestic heating market is largely supplied by the latter. At present the gas companies do not see any great change in this pattern, although the Japanese consumer has become much more oriented towards space heating. Because of the short extent of very cold weather in winter on the south coast of Honshu, supplies to the domestic heating market would also have a very poor load factor. It is apparent, therefore, that LNG supplied in the 1970s will not be able to rely on this high value market to the same extent as in the U.S.A.

Japan, however, like the U.S.A., is energy-short and suffers from very severe atmospheric and other pollution problems. Natural gas as a sulphur-free fuel has great attractions for those large fuel users who are, in agreement with local authorities, forced to use very low average sulphur content fuels. Among such fuel users are the power generation and iron and steel industries, who between them are responsible for a very large portion of total sulphurous emissions in Tokyo, Osaka, and other highly industrialized areas. Tokyo Electric are operating power stations on Alaskan LNG and have contracted with Shell/Mitsubishi already for supplies from Brunei for use in new power stations. Supplies already contracted are competitive with other forms of very low sulphur fuel. Future potential supplies from the Persian Gulk, Sarawak, Australia, and U.S.S.R., if purchased, will be no doubt largely for use as low sulphur fuel and will have to compete on cost with other methods of reducing sulphurous emissions.

The iron and steel and power generation industries, as well as local authorities and MITI, are examining carefully all possible means of reducing sulphurous emissions to the agreed limits.

LNG has to compete in the very low sulphur fuels market ( $< 0.5$ per cent) with:

Low sulphur Indonesian crudes and residues
Naphtha and condensates
Refinery and field-produced LPG
Distillates

and for demand for 0.5-1.0 per cent sulphur content fuel with:

Fuel oil blends
Desulphurized fuel oil.

None of these possibilities alone can offer the complete answer to Japan's demand problems. Flue gas desulphurization may be attractive for large users, as it allows them to retain some measure of flexibility in their forward fuel supply planning. Costs for this route have been variously estimated at 12-26 cents/$10^6$ Btu, depending on plant size for 90 per cent sulphur extraction using 3 per cent sulphur fuel oil.

Much development work is proceeding in Japan on processes for stack gas washing and it is likely to be an important contributor to pollution reduction in the 1980s. Certain processes, however, cause waste disposal problems which have not yet been adequately resolved. Stack gas washing would be, however, an uneconomic route for certain industries such as iron and steel. The cost of desulphurization by this route also indicates the size of the sulphur premium power generation companies and other large users might be willing to pay for purchasing low sulphur fuels.

The desulphurization of fuel oil down to 1 per cent sulphur content appears to be practical. Estimates of the order of 5-6 cents/$10^6$ Btu for each 1 per cent sulphur removed down to 1 per cent have been given. Costs may rise steeply, depending on the type of crude, if removal of sulphur to levels below 1 per cent is required. Fuel oil desulphurized this way would have to be further blended to achieve a further reduction in fuel sulphur content down to the levels necessary for the very large users.

Other major competitors to LNG are naphtha and Indonesian crudes and residues. Both are to some extent supply-limited, and naphtha imports for burning may be restricted by pressure from the petrochemical industry keen to protect its own feedstock availability and prices. Some authorities think, however, that naphtha will eventually be imported for burning in quite large quantities.

The sale of LNG in bulk as a low-sulphur fuel may not be profitable for the supplier, since many of the premium qualities of natural gas are not required and remain unvalued in the price, and for a reasonable value LNG sellers may rely in the long run on growth in the premium industrial process and domestic heating markets. Such demand, it must be stressed, is unlikely to form the major component of schemes planned for deliveries in the 1970s.

## West European Markets

Although Western Europe receives natural gas from two out of the four existing LNG shipping schemes, there is less immediate need for large quantities of base load gas at present than in Japan or the U.S.A. This is because the present schemes result from demands which are atypical of the current European markets. The Canvey Island-Le Havre scheme ($53 \times 10^9$ ft$^3$/year) from Arzew was originally based on the competitiveness of LNG with town gas manufactured from naphtha in the early 1960s. Since the discoveries in the North Sea, the United Kingdom gas demand is satisfied by cheaper piped gas from that source. The LNG at Canvey Island therefore acts at present as a small supplement to base load supplies and also as a useful source of peak-shaving gas.

The supply of LNG from Marsa el Brega to Barcelona ($39 \times 10^9$ ft$^3$/year) meets a demand for gas to be distributed in the domestic and industrial market at a lower cost than manufactured gas, since Spain has no commercial natural gas production. The supply to La Spezia ($85 \times 10^9$ ft$^3$/year) acts as a supplement to the production from indigenous Italian fields, which has reached plateau levels, allowing existing investment in mains and distribution systems to be more fully utilized. The short shipping distance across the Mediterranean also means that the landed cost of LNG to these buyers (43 and 39 cents/$10^6$ Btu respectively) is competitive with natural gas from Northern Europe and the U.S.S.R.

For Northern Europe as a whole, however, the huge gas discoveries at Groningen and the North Sea mean that there is no present shortage of natural gas for LNG to supplement. Major European governments at present have not yet laid down very severe restrictions on the sulphur content of fuels used. But as legislation reducing sulphur concentrations in the air is put on the books, natural gas will increasingly obtain a premium for its low sulphur qualities.

Increasing demand in the domestic heating and industrial process markets, as well as some growth in the high load factor, low value under-boiler market, is likely to lead to a need for supplemental sources of natural gas by the late 1970s and some analysts see a need for $700$-$1000 \times 10^9$ ft$^3$/year at that time, of which some 60-65 per cent will be supplied by pipelines from the U.S.S.R. to Germany, France, Italy, Austria and Finland, and some 35-40 per cent coming from North Africa through existing schemes and to Fos from Skikda. Thereafter, further supplies of natural gas will become necessary. Possible sources are by pipeline from the U.S.S.R. and the North Sea, LNG from Algeria and Nigeria, and the manufacture of SNG from naphtha or crude oil. In addition, there are the possibilities, not yet fully evaluated, of pipelines under the Mediterranean from North Africa.

The forecasts referred to above assume that natural gas will remain competitive with fuel oil for the under-boiler market. Before European gas marketers commit themselves to taking supplies of high load factor, expensive supplemental gas, they will have to be satisfied as to the economics of continuing to supply the low-value under-boiler market, and here the cost of fuel oil will be critical.

In the U.K., and ultimately in Continental Europe, the prospects for imports of LNG depend upon the continued development of home heating demand and the extent of future discoveries offshore.

## PEAK-SHAVING

Gas marketers have to provide the system capacity and the product to meet extreme peaks of demand. The costs of supplying year-round demand differ from those of supplying seasonal and peak demand. Since gas marketers have to improve the system load factors to reduce unit costs, they have to examine whether it is more economical to sell interruptible gas in off-peak seasons or to supply low load factor demand from a different and usually less capital-intensive source per unit of peak demand, or to combine both approaches.

Possible sources of peak and seasonal supplies include:

(i)      Underground storage in depleted oil or gas fields or in aquifers.

(ii)     The manufacture of partly or fully compatible substitute natural gas.

(iii)    Storage of LNG under or above ground.

Underground storage of natural gas is normally more economical for supplying seasonal rather than peak demand, since the very high rate of production necessary to meet needle peaks would require a very high investment in compression equipment and production wells. The possibility of using underground storage for seasonal demand must also depend upon the existence of depleted oil or gas fields or other suitable structures, such as aquifers, sufficiently close to the point of demand. Generalizations about costs of storage by this method are not possible, but factors affecting cost are the permeability and depth of the structure which affect the number of wells to be drilled and compression necessary, and the distance from the demand area and the size of available storage which affects the length and diameter of the pipe.

LPG-air systems are excellent for needle peaking, as they can be highly flexible, easily automated, and can be brought on line within an hour, as well as having low fixed costs which compensate for high feedstock costs. LPG-air mixtures have to be introduced into the main natural gas supply in strictly controlled proportions owing to differences in combustion characteristics. In the U.K. it has been calculated that by restricting the LPG-air mixture to 10 per cent of gas delivered, the combustion characteristics remain acceptable while allowing gas of the correct calorific value to be distributed thus avoiding any loss of revenue through giving away heat. An LPG-air system at the demand end of one such transmission line has led to an improvement in load factor from 40 to 60 per cent and deferred the expenditure on a new pipeline for several years.

Manufacture of SNG from naphtha for peak-shaving may be economic where an old fully-amortized manufactured gas plant exists, when production costs may be largely feedstock and operating costs. But for peak-shaving the process must allow for rapid start-up times from cold or standby, and the costs of keeping the plant on standby have to be taken into account.

The popularity of LNG for peak shaving is increased by the development of demand for low load factor domestic heating which, in the more northerly parts of the U.S.A., U.K., and Europe cannot be offset by similar increases in summer demand for air conditioning. LNG is capable of very high volumes of send-out at very short notice and can offer security of supply at the point of peak demand to the distributor if other sources fail. This location at the point of demand favours LNG against underground storage. The necessity for the injection of cushion gas into new underground storage at a time of actual or potential shortage of indigenous gas also favours the use of LNG.

There is increasing scarcity of further structures suitable for underground storage near to major centres of demand, which could be economically operated.

Of importance to the gas marketer will be the increasing availability of LNG, either from plants built to liquefy indigenous pipeline gas or from very large export plants overseas. If a marketer has a sufficiently large annual peaking capacity requirement, it may pay him to build his own plant to liquefy gas during valleys in demand, thus using otherwise unused pipeline capacity. Otherwise he may consider purchasing from a larger plant and merely operating satellite storage.

For instance, it has been calculated that ten separate gas utilities, each with a requirement of $200 \times 10^6$ ft$^3$ of annual peaking capacity, could reduce their ex-liquefaction plant unit cost of LNG by 65 per cent if they were to purchase from a plant sized to meet their total demand. The final economics of such a scheme would, of course, depend upon the distance of the utilities from the joint plant and the resulting transportation costs. The scheme illustrates, however, the flexibility in planning which LNG can offer.

The nature of peak-shaving makes the value of the various peak-shaving methods dependent upon the peculiarities of local demand patterns and supply capabilities. The value of LNG for this use will therefore vary widely in the major markets of the world.

## LNG AS AN AUTOMOTIVE FUEL

Increasing availability of LNG in the U.S.A. from the development of LNG peak-shaving plants throughout the country, as well as the possibility of large

base load imports, has stimulated interest in its characteristics as an automotive fuel, especially for the reduction of noxious vehicle engine emissions.
In the U.S.A. gasoline and automotive diesel engines presently in use are
creating serious atmospheric and noise pollution problems. So far, the
emissions from vehicle engines that have caused most concern have been
carbon monoxide and unburnt hydrocarbons and nitrogen oxides. There is
also a desire to reduce noise from diesel engines.

The use of LNG or LPG (though supply-limited) in place of gasoline and
automotive diesel fuel can make it possible to reduce all these pollutants. In
addition, by using the cold in LNG and a high compression ratio (12:1),
the power output may be kept close to that of gasoline, if lean mixture
operations are compared.

The leaner running and higher compression ratio of an LNG engine results
in decreased specific consumption relative to gasoline, and to a lesser extent
relative to butane. Development of low pollution engines to meet the 1975-76
Muskie limits will lead to the increased complication and cost of gasoline
engines. To date, neither LNG nor gasoline have achieved the very low $NO_x$
levels required. But it may be expected that the lean running LNG engine will
have lower maintenance costs and a lower specific consumption for similar
capital costs per vehicle.

LNG is unlikely to be injected as a liquid fuel into the combustion
chamber of a diesel engine because of the very low cetane number and because
its temperature would promote severe injector design problems. Spark ignition
is therefore more likely. The heavy construction of the current diesel engine
could be used to make full use of the high anti-knock rating and cold of LNG
to increase the power output by turbo-charging and intercooling. The value
of this power increase would vary considerably from one application to
another.

Of great influence on the possible price of LNG to the user will be the tax
imposed by governments upon LNG relative to gasoline. If LNG's anti-
pollution possibilities cause governments to charge tax at less than thermal
parity with gasoline, LNG use will be encouraged. Equal taxation on a
volumetric basis would, however, be detrimental to LNG's chances, as the
energy content per unit of volume is only 67 per cent of that of gasoline.

In view of the heavy capital investment that would be necessary to install
a full distribution system for LNG, the most likely future for LNG is as a fuel
for some commercial vehicles fleets driven in one locality and garaged and
serviced at a central point. In this way distribution costs are minimized while
city centre pollution emissions are reduced.

## UTILIZATION OF COLD IN LNG

LNG marketers face in each potential sale of LNG the challenge of
obtaining a value for the cold in LNG additional to the value of the gas.

Technologically, however, it is not possible to recover more than a fraction of the theoretical (Carnot) work introduced during liquefaction at the re-vaporization.

Although the technical considerations in using the cold are relatively straightforward, it is not easy to develop an overall scheme, such as is planned for Fos-sur-Mer, comprising an industrial complex in which one component factory has to rely upon another for energy, products, or cold, and the development of Fos has taken many years and is yet far from completion.

Manufacture of liquid oxygen and nitrogen is a good example of the use of cold at very low temperatures (knows as "deep cold"), with consequent benefit in overall thermodynamic efficiency. This use of cold can reduce manufacturing costs of liquid oxygen and nitrogen by 20 to 30 per cent if the cold is assumed free of charge. The problem is that if the total amount of cold avilable from a terminal were used to produce oxygen and nitrogen, production would be far in excess of present market demand, and so only a small proportion can be used.

To obtain the greatest value from LNG cold, the maximum use of very low temperatures is desirable but as yet the application of cryogenics is in its infancy. In many cases the use of an intermediate cold carrier will be necessary. It would be important to develop carefully around an LNG terminal a market for varying grades of cold in order to maximize benefits potentially available. Even so, the contribution obtainable though significant to overall economics is not likely to influence potential buyers' choice of fuels greatly.

## CONCLUSION

From the foregoing it can be seen that the full value of LNG is most likely to be obtained when liquefaction allows natural gas to be supplied for premium uses such as climate conditions or high value industrial process uses (*i.e.* metallurgical, ceramic, food, and chemical industries). The competition for these uses comes from other forms of gas making or additional supplies of natural gas. Further into the future the use of LNG as an automotive fuel may provide further high value markets, while the development of cryogenics may in time provide a significantly larger contribution than present technology allows.

## BIBLIOGRAPHY

Clarke, D. J., Cribbs, G. S., and Walters, W. J., "The Philosophy of Gas Storage", Inst. Gas Eng.

Biederman, N. P., "Satellite Peak-shaving with LNG", Pipeline & Gas 1970.

## DISCUSSION

*J. L. L. Orbach* (Northern & Central Gas Corpn Ltd) congratulated the authors on their very interesting paper. The figures for the shortfall of supply in the U.S.A. were based on two reports by the Committee of Future

Requirements and the Committee of Future Supply. Both these reports were misleading as guides to the real energy picture, in that their conclusions were based on the assumption of a *status quo* with regard to prices and other conditions. Already indications were that prices were altering to reflect the situation which was developing

In many ways it was more realistic to supply the energy gap in the U.S.A. by pipeline from Canada rather than by LNG shipments.

*I. A. I. Griffiths* felt that LNG would not necessarily be the preferred fuel and that the U.S.A. would need to make use of all available energy sources, with economics being the determining factor in which type of energy would supply the greater portion of the market. National security could lead to a premium being placed on pipeline gas from Alaska but only time would tell how significant such a premium would be.

*Miss M. P. Doyle* (Esso Petroleum Co. Ltd) questioned the realism of assuming that the potentially available reserves in the U.S.A. would be developed to fill the energy shortfall because present low wellhead prices did not justify the cost of doing this.

She felt that in order to establish a higher price for natural gas, more effort should be directed at selling to the premium market, such as commercial usage for oil conditioning, etc., which by U.S. experience could provide an important market.

*I. A. I. Griffiths* agreed that commercial usage offered good prospects for additional sales in the U.K. of which full advantage had not yet been taken. He felt that figures produced by the Gas Council for this section of the market were conservative.

*R. L. Torczon* (Conoco Europe Ltd) said that the cited U.S. reports were unrealistic simply because, in order to arrive at some conclusions, basic parameters had to be established. One of the assumptions was that economic factors would remain stable.

*A. R. Khan* (Gas Development Corpn) agreed with Mr Griffiths' assessment of the supply and demand picture in the U.S.A. He felt that the U.S.A. should import more of its gas requirement from Canada in preference to LNG importations from politically unsuitable areas. The decision on these imports, however, rested with the U.S. Energy Board.

The scale of the problem was illustrated by the fact that in order to maintain the existing Reserve Life Index it would be necessary to discover at least 20 trillion cu ft of gas reserves every year in the future.

*M. W. H. Peebles* (Shell International Gas Ltd) felt that it was a mistake to review the future of gas in isolation. The more important issue was the overall energy demand throughout the world and the position of gas should be viewed within this context.

The importance of the world energy picture could best be illustrated by the fact that the oil industry would need to find as much oil in the next ten years as it had done in the last 100 years if the demand was to be met.

Looking at the overall energy availability, much had been said about a growing new requirement in the U.S.A. for coal obtained by strip mining, but the environmental effect of strip mining on the scale necessary would be far more serious than the effects of importing the shortfall of energy in the form of LNG.

# Conversion to Natural Gas in the Public Distribution Sector

## By H. J. BUCKLEY

### *(Imperial Continental Gas Association)*

## INTRODUCTION

The gas industry in Belgium, as in other countries, has experienced during its long history periods of growth and periods of comparative inactivity. Nevertheless, the resilience of the industry is evident in that, after 150 years, it is successfully marketing much the same product as it did when it began, despite the fact that in its lifetime it has undergone a number of basic changes in production technology and customer requirement.

With the advent of natural gas, the industry has had to face its biggest challenge to date and develop new methods and techniques to cater for the new situation. The time is well chosen, therefore, to paint a picture of gas distribution in Belgium, which appears to be fairly typical for Europe, and to present the various facets of the problem as fully as possible within the space permitted.

The first part of the paper sets out the evolution of the industry up to the time of the arrival of natural gas, the second part attempts to define the economic conditions which govern the distribution of natural gas and the third part draws certain conclusions about the present situation, with particular reference to our Group's experience

## THE DEVELOPMENT OF GAS SUPPLIES IN BELGIUM

It is customary to divide the gas industry into three functions, production, transportation, and distribution, which are nevertheless interdependent.

### Production

Coal gas was first distributed through mains as far back as 1814 by gas industry pioneers such as Murdock, Minckeleers, and Lebon. Originally used for public lighting, it later became a domestic substitute for oil lamps.

At the end of the last century, these applications were successfully challenged by electricity. This meant that the industry had to find new sales outlets; therefore, around 1900 it penetrated the cooking market and by 1910 it had extended its activities to water heating, industrial applications, and,tentatively, space heating.

The need of the steel industry for coke also dates from about 1910 and led to the building, often at the steelworks, of coke ovens which produced gas as a by-product. Between 1920 and the beginning of World War II, there was a surplus of coke oven gas and distributors were able to consolidate their markets and develop base load space heating applications. After the war, a depression in the steel industry and new technology in steel production limited the production of coke and hence the availability of coke oven gas.

The gas industry was obliged, therefore, to develop other methods of gas production and by 1950 reformed gases based on feedstocks of methane (from coal mines), LPG, light distillates, and refinery tail gases had become competitive. Autothermic reforming plants for LPG feedstocks were installed alongside coke ovens, and at the same time cyclic reforming plants for cracking large quantities of LPG or light distillates were built on separate sites. Gas production was becoming independent of coke production and distributors developed their sales, not only because gas had become more competitive, but because production could more readily follow seasonal demand.

The availability of natural gas from Algeria in 1959, and later from Holland, prompted distributors to launch space heating campaigns which they felt were necessary to take full advantage of the natural gas when it became available, but this accelerated the tendency towards a reduction in the regularity of demand.

Fig.1 shows that the difference between minimum and maximum outputs had been increasing exponentially, which gave rise to increasingly heavy annual investments in production plant. Over a period of 30 years to 1965, peak demands had increased sixfold, but annual sales only trebled.

The further acceleration of this tendency as a result of the sales initiative taken by the Belgian gas industry could only have been envisaged in expectation of natural gas supplies, and explains why distributors started planning their conversion as a matter of urgency by 1963. The Slochteren field was so near at hand that further major investment in reforming plant seemed illogical.

## TRANSPORTATION

Around 1930, with the increasing distances between the points of production and the points of consumption, welded steel mains became necessary in order to carry the gas at 1-4 Bar. (15-60 psi), then considered to be a high pressure. This marked the beginning of the differentiation between production, transportation, and distribution and led to a rapid expansion of the

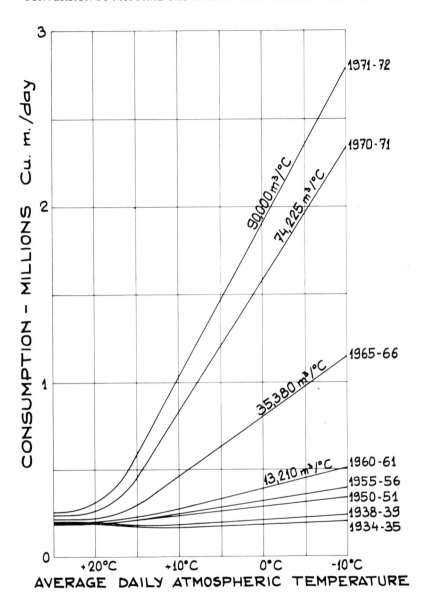

Fig.1 Daily consumption relative to atmospheric temperature

transportation function, under separate management. Transporters had to resort to compression of the gas to improve the coefficient of utilization of their pipelines.

Thereafter, new production plants, based on LPG and refinery tail gases, were built near the centres of consumption, since it was preferable to carry the rich feedstocks to the works and shorten the gas distribution networks. Predictably, the feedstock was brought in by pipe, if gaseous, and by ship, barge, and lorry, if liquid, and the two were combined to maximize the dependability of gas supplies and minimize the cost.

At the time when natural gas was introduced into Belgium, transportation pipelines were being designed for 5-15 Bar. (75-220 psi), but since then a national grid system designed for 67.5 Bar. (1000 psi) has been superimposed, in view of a guaranteed 70-Bar. at the frontier.

### Distribution

A distribution network operated by a local distributor comprises a system of mains to carry gas away from an urban centre of supply which can be either a production works or a regulator station. This feeds a series of mains and service connections to reach all consumers (see Fig. 2).

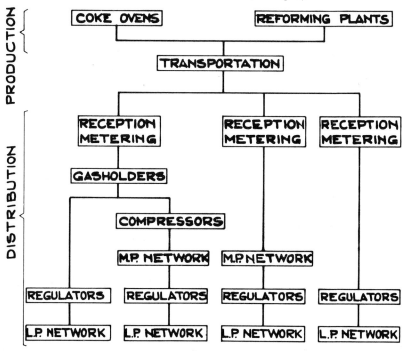

Fig.2

During the "gasworks" era, this was done exclusively at low pressures, which meant that the mains had to be of relatively large diameter. Gasholders often held a whole day's supply to allow for periodic shut-downs of the production plant.

The introduction of link mains between supply centres reduced the need for storage to the equivalent of about eight hours production and finally the introduction of cyclic reforming plants superimposed on other means of peak production made further modulation possible and reduced the need for storage still further.

Originally, distribution mains were made of cast iron, which has adequate resistance to both mechanical stresses and corrosion. Joints were mainly of the socket and spigot types, but various methods were used to seal them (see Fig.3) ranging from jute and lead seals to rubber sealing rings, whilst various types of cramps to compress specially shaped sealing rings were also in use. These systems gave little trouble until the introduction of long-distance transmission of gas which, because it was relatively dry and had also been stripped of its benzole, caused the jointing materials to dry out and shrink. This led to leakage.

When asbestos cement pipes came on the market, they were selectively introduced because suitable grades were available at a lower price than cast iron.

In distribution networks, steel mains with welded joints came into their own, particularly for higher pressures, but they are prone to corrosion and require cathodic protection. This became a matter of routine, but marrying sections of steel main into existing cast iron networks created its problems.

As a result of these developments, the networks, just before the introduction of natural gas, consisted of different types and ages of main in proportions which varied by districts. They included joints made of jute and lead, natural or synthetic rubber, mechanical joints, and those of welded steel construction.

## THE ECONOMICS OF CONVERTING GAS DISTRIBUTION NETWORKS TO NATURAL GAS

### Problems Posed by the Discovery of Natural Gas

The discovery of natural gas poses complicated economic problems to distributors, and it cannot be assumed that in every case it will pay a distributor to take a supply of natural gas.

The geographical location of the network relative to the source of supply must be such that the cost of transporting the gas is not prohibitive. In Belgium the distance of the major towns from the Slochteren field in Holland is only of the order of 200 miles, and there was no doubt about feasibility by pipeline, so the study required was a comparison of the economics of using natural gas supplied by pipeline with those of continuing to use

## CAST IRON PIPES

JUTE AND LEAD JOINT

"TRIFET" SYSTEM
WITH RUBBER SEALING RING

"UNION" SYSTEM
SCREWED MECHANICAL JOINT

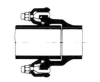

"PRECIS EXPRESS" SYSTEM
FOR DUCTILE IRON PIPES

## ASBESTOS CEMENT PIPES

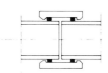

"SIMPLEX" SYSTEM
SLEEVED JOINT

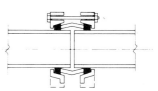

"GIBAULT" SYSTEM
BOLTED JOINT

## STEEL PIPES

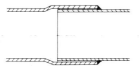

SLIP JOINT WELDED

BUTT WELDED JOINT

Fig.3 Different types of joints

manufactured gas produced locally by reforming liquid or gaseous hydrocarbons, and from coke ovens.

In each case the comparison of costs had to be made by establishing a discounted cash flow (DCF) value for every feasible hypothesis, having regard to the cost of the raw materials, labour costs, capital investments, the tariff established for the purchase of gas from the bulk transporter, and, in particular, its sensitivity to variations in demand, the cost of the supply main, the cost of conversion, the size of distribution operation, the rate of increase of consumption, the evolution of the load factor, and the discount rate adopted for calculating DCF values. It had to be ascertained whether the natural gas was available in sufficient quantities at a suitable price and with a sufficient guarantee of deliverability, also whether it would be possible to place additional volumes of gas. In our case the results appeared favourable.

Once it was decided to use natural gas, it was necessary to determine how best to do this to achieve optimum economy:

— Either to reform the natural gas from 8400 kCal/m$^3$ (944 Btu/cu ft), into a town gas substitute at 4250 kCal/m$^3$ (478Btu/cu ft), leaving the distribution networks and consumers' appliances as they were.

— Or to introduce natural gas into the network progressively, as the transportation and reception equipment, the distribution network, and the consumers' appliances were adapted to receive it.

The choice of one of the above solutions was based on an evaluation of their economic consequences. They are:

— Reforming the natural gas entails continued investment in reforming plants to match consumption. These give rise to operating costs, high maintenance charges, and permanent process losses, as well as to heavy amortization and financial charges which are independent of utilization. Furthermore, they have a limited capacity which restricts the availability of supply. On the other hand, this solution avoids conversion of the networks and consumers' appliances.

— Natural gas distribution necessitates new expenditure comprising, for example, the adaptation of the distributors' regulators, meters, and valves, basic modifications and repairs to the existing distribution network, and, particularly, the conversion of consumers' gas appliances. On the other hand, the thermal carrying capacity of the network is increased by some 75 per cent and the relatively high gas pressure at city gate makes it possible to dispense with compressors.

Provided that the necessary finance were available, the second solution, that of conversion to natural gas, appeared in our case to be more advantageous in the long term.

### Problems Posed by the Arrival of Natural Gas

Once the decision is taken to distribute natural gas, the technico-economic problems of the conversion itself have to be resolved. These involve a very large number of variables and a fresh study is required to consider a breakdown of such items as general conversion policy, the preparation of conversion techniques, financial administration, commercial policy, etc. These are interdependent, but for convenience they are discussed separately below.

## General Conversion Policy – Organization

One of the first problems to be solved is timing, which means the choice of starting and finishing dates, to the mutual satisfaction of the distributor and his consumers. Rate optimization is achieved by a compromise between technical, commercial, and financial considerations, *i.e.* the art of the possible.

It is necessary to find, train, equip and organize the conversion teams, organize stores, workshops and back-up services, and plan the entire conversion operation in minute detail. In practice, the law of diminishing returns sets a top limit to the number of consumers converted per week in any given case, and the weekly sectors can vary considerably, *e.g.* from 1500 to 4000 consumers per sector.

## Technical Problems

Conversion poses numerous technical problems, relating basically to the conversion of gas appliances, and to the distribution network.

## Gas Appliances

It is necessary to analyse the gas appliances in use and select the optimum gas pressure, the basic operations required to execute the conversion, the methods to be used to adapt burners, the tooling, and the materials required.

In the case of Slochteren gas, which contains 16 per cent non-combustible gases, the pressure required at the appliance over a range of burners tested had theoretically to be about 18 millibar (7.2″ WG), but in practice the less efficient burners required more than this, and since Holland had previously adopted 25 millibar (10″ WG) and appliances were already available for these conditions, Belgium decided to adopt the same pressure. The practical limits of pressure variations in the networks were set at 20-30 millibar (8-12″ WG), against 8-15 millibar (3.6″ WG) for town gas. This involved a twofold or threefold increase in the low pressure (LP) distribution pressure and led to a substantial increase in gas losses by leakage from the networks.

Having set the distribution pressure, the conversion methods had to be chosen when planning the conversion of an appliance. In order to do this, it was necessary to consider the following criteria:

| | | |
|---|---|---|
| Technical | – | Can it be done without damaging the appliance? |
| Safety | – | Will the converted appliance be safe to use? |
| Output | – | Will the thermal output be satisfactory? |
| Financial | – | Will the cost of conversion be acceptable? |

To give an idea of the magnitude of the problem, there were over 10,000 variations in appliances in Belgium and solutions to the majority of these had to be found within six months. However, the conversion can be simplified by persuading consumers to buy new appliances fitted with multi-gas burners at special discount rates, in exchange for their old appliances. The discount was set against what would have been the cost of conversion. The "multi-gas" sales campaign, while continuous at a national level, was also focused progressively on areas due for conversion a few months later.

A few appliances could not be converted within the limitation set out above, but for the majority there was a choice of methods of conversion. The particular problem with natural gas is to find a method of stabilizing the flame, but, unfortunately, no universal solution has been found. A certain number of appliances were converted by standard replacement burners or "sets", supplied by the appliance manufacturers, but the remainder were converted by various other methods. The economy in labour by using replacement burners resulted in greater conversion speeds with the available workforce and, as had been anticipated, this economy more than offset the higher material costs. This led us to develop and produce our own sets in many cases. Supplementary advantages were minimum disturbance of consumers, the possibility of converting a wider range of appliances, and the modernization of existing appliances.

Looking back we are satisfied that the choice of conversion methods was essentially correct within the limitations of current knowledge and availability of materials, but nothing is ever perfect and experience has shown that:
— The burners of multi-gas appliances, while serving a useful purpose during conversion, were a compromise and some are less efficient on natural gas than burners designed specifically for this gas.
— The consumption of pilot flames doubled.
— A large number of appliances were less responsive to regulation than they had been on town gas and no real solution was found to stabilizing very low flames.

### Distribution Installations

Apart from the physical effect of natural gas in gas networks, which is mentioned later, conversion has a basic effect on the network because of:
— The building of new reception stations to suit the new trunk supply mains.
— The transformation of existing reception stations to take inlet pressures of 15 Bar (220 psi), instead of 8 Bar (115 psi), and higher outlet pressures both into the medium-pressure network working at 5 Bar (75 psi), instead of 0.5 Bar (7 psi), and low-pressure network, 25 millibar (10″ WG) instead of 10 millibar (4″ WG).
— The adaptation of certain regulators to work on the new LP regime and the replacement of regulators expected to receive 5 Bar (75 psi).

— The division of the network into sections or sectors which could be isolated for the turn-in of natural gas and the conversion of appliances. This was usually done in sections representing one week's work.
— The higher distribution pressure and dryness of natural gas which causes an increase in leakage, making it necessary to perform a leakage survey and to repair the leaks found in the whole network ahead of conversion, followed by a repetition of this process after conversion.
— Certain types of gas meters proved to be unsuitable and had to be replaced.

## Financial Problems

The methods adopted for financing conversion varied considerably between countries, and the definition of conversion cost also varied.

Generally speaking, in our networks the gas distributor bore the cost of modifying the network and converting the appliances of domestic and commercial consumers. Appliances declared unconvertible were the subject of special discounts for replacements and involved the appropriate commercial action.

Industrial consumers, whose equipment was often costly to convert, were invited to pay for the conversion themselves, but in return they were offered a lower gas tariff to compensate for this expense. This had the further advantage of encouraging future consumption of gas.

Conversion costs of domestic consumers varied with geographic location, the density of population, and the average number of appliances per consumer, but generally fell between Frs.B.4000 and 5000 (£34-43) per consumer. This refers to Belgian wage rates etc. and included an apportionment of overheads and amortization of temporary equipment, also some work on networks, including sectionalizing, meter replacements, locating and repairing leaks, etc. Some companies allocated some of these costs to other accounts, which reduced their apparent conversion costs.

In Belgium a fund was created on a national basis, under the aegis of the Gas Industry Federation, to supply loans for conversion at the rate of Frs.B. 4000 (£34) per consumer, but as this sum did not cover the whole cost, the balance had to be found by the distributors themselves.

The reimbursement of these loans, plus interest, is effected by a payment of approx.15 per cent of the cost of gas purchased by the distribution companies which, together with a 5 per cent contribution from the pipeline company, was calculated to close the account in 20 years.

## Commercial Problems

The price of natural gas basically fixes its position in the energy market and although the gas industry in Belgium is not nationalized, public authorities were involved in the negotiations for the frontier price paid by the pipeline company, as well as the city gate price paid by distribution companies. Consumer prices are also controlled by a national committee.

At the time when natural gas was introduced into Belgium, the Belgian State took a 33.1/3 per cent participation in the pipeline company, having given this company a virtual monopoly of supply to industrial consumers which used more than 8000 Gcal/year (approx.1 million m³/year). In view of their better utilization factor, the gas is normally sold to these consumers at a lower price than to the distributors, who previously had the monopoly of gas supplies in their areas. This led to a rapid development of industrial sales, which now account for 80 per cent of total sales in Belgium.

This situation has had a profound effect on distributors, who must face certain fundamental facts:

— The economics of conversion are based on a rapid expansion of gas sales to cover the high cost of the operation.
— Annual charges on total investments, old and new, together with operating overheads, represent a total cost which is practically independent of gas sales.
— The necessary sales of gas must be effected in the heating market which, by its very nature, leads to poor utilization factors. Moreover, it is in direct competition with other fuels.

The tariff policy of distributors had to take account of these facts and meet two fundamental requirements which are to some extent contradictory, *i.e.* to maintain overall profitability whilst at the same time offering competitive prices.

Bearing these points in mind, the action taken on conversion was, first, to change town gas tariffs into natural gas tariffs, with a 5 per cent reduction in price on a thermal basis, so that consumers would not suffer any increase in their gas bills for an equivalent service. Secondly, new tariffs were introduced to increase penetration in the space heating market in certain categories of consumers, *e.g.* blocks of flats, commercial premises, and large houses.

Right from the outset, however, these tariffs were regarded as provisional and subsequent reductions in the purchase price of gas were passed on to consumers, with a result that the domestic heating tariffs for natural gas, having regard to consumption, can show reductions of 10-20 per cent, even 40 per cent in a few cases, on the old price of town gas. Nevertheless, the increase in gas sales has not come up to expectations.

In 1966, when the gas purchasing contracts were fixed on a national basis, they set an objective of 3000 million m³/year to distributors by 1975, *i.e.* a quadrupling of sales in ten years or about 15 per cent/year. This rate may be achieved later, but to date distributors have only reached about 11 per cent/ year, which reflects some success in replacing coal for heating and also a general increase in fuel requirements, but current gas prices in Belgium make it difficult for distributors to compete with oil in the heating market.

## After Conversion

The evolution of gas distribution in the years immediately following conversion is conditioned by:
- The specific characteristics of the gas itself — high calorific value, available pressure, dryness, purity.
- The conditions of contract for its purchase.
- Its development potential, represented mainly by the heating market.
- A constant effort towards greater safety.

These considerations affected the form of distribution, the materials used, the management of the network, and the marketing of the gas. These aspects are reviewed below:

## Evolution of Distribution Control

As pressure considerations have become critical, it has become increasingly necessary to install remote indication of pressures at key points. It is now intended to have a central monitoring desk with a high degree of remote control of gas pressures. In case of accidents, this will enable the operator to take immediate action, as well as alerting the mobile service squads by radio-telephone.

## Smoothing of Hourly Consumptions

An effect of selling in the heating market is to increase the morning peak consumption, which has now caught up the noon cooking peak. This has a favourable effect on the hourly utilization factor, as shown in Fig.4, and it is hoped in due course to cease using gasholders to meet hourly peak demands, and eventually to be able to scrap them.

## Smoothing of Daily Consumptions

The daily utilization factor, on the contrary, is adversely affected by the increase in heating load. Efforts are being made to improve this situation by finding summer loads, but the pipeline company's effective monopoly of major industrial loads blocks this market for distributors. Furthermore, the Belgian climate does not favour air conditioning and distributors cannot economically acquire interruptible loads. Belgian distributors' purchasing contracts with the pipeline company also preclude the manufacture of substitute gas to reduce peak day intakes from the gas trunk lines.

A clause in the contract sets a top limit to the price payable by distributors, but the daily peak situation nevertheless has an increasingly important effect, and this problem is receiving considerable attention. Liquefaction off-peak may be a solution in due course, but appears to be uneconomic at the present time.

## The Secondary Transport Network

The development of the medium-pressure system is conditioned by the following considerations:

— The necessity to meet increasing peak demands will involve its reinforcement as well as the multiplication of district regulators and consumers' regulator cabins.

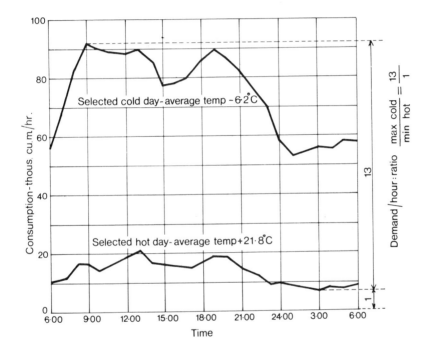

Fig.4 Hourly consumptions

— The necessity to interlink the supply points in order to rationalize flows and pressures.
— The availability of 15 Bar (220 psi) pressure at city gate, which permits the adoption of this pressure out of town, and of 5 Bar (75 psi) in town. Their adoption necessitates the installation of the appropriate valves, regulators, etc., to withstand these pressures and suitable safety equipment is also being adopted as standard.

— Certain sections of network which were acceptable for manufactured gas are not acceptable for natural gas, and have to be modified, down-graded, or scrapped.
— Networks are being increasingly interlinked and fitted with valves at key points to control the gas while permitting continuity of supply in adjacent districts.
— New mains are of welded steel pipe, which has to be suitably coated and cathodically protected.

## Local Distribution Networks

The following considerations apply to the low-pressure system:
— The availability of 5 Bar (75 psi) pressure makes it possible to supply rural areas through small diameter mains, which makes a gas supply to certain areas economically viable when this was not so before. Furthermore, larger consumers are increasingly supplied at this pressure.
— Where practicable, humidification of the gas is practised to prevent the jute in the lead-jointed cast iron mains from drying out and leaking. Moreover, the high level of activity in repairing leaks will continue for some years so as to condition the network to the new gas as quickly as possible. New methods of sealing mains and services are being tried, with varying degrees of success, but where leakage is high, mains are being replaced systematically.
— The replacement of meters must be accelerated because certain models are technically unsuitable for use with natural gas, because of the higher pressures.

## Legislation

A law has been passed in Belgium governing in detail the laying of high-pressure (over 15 Bar) trunk lines. A second law followed covering safety measures to be adopted in the distribution of the gas to consumers.

## Evolution of Materials Used

As regards mains, for pressures over 5 Bar only welded steel is used. For low pressures (0.1 Bar) and medium pressures (up to 5 Bar), steel and asbestos-cement will continue to be used, whilst ductile iron and plastics will be used increasingly.

As already mentioned, regulators and associated equipment must be designed for higher pressures. This, and particularly the safety equipment which must now include a safety cut-off valve, is more costly than that previously used.

Meter sizes are being related to throughput and made with more robust cases which are less likely to suffer mechanical damage.

## Evolution of Network Management

Network capacities must match heating demand, whose peak is increasing exponentially and is highly sensitive to atmospheric conditions. It is important, therefore, to plan reinforcement mains well ahead of requirements because of (a) the time required to plan the route in detail in an already congested sub-surface area, (b) obtaining planning permission from the appropriate authorities, (c) getting delivery of the pipes, and (d) employing a contractor to lay them, quite apart from the time taken for the actual laying.

Planning entails statistical forecasting of consumption combined with a detailed analysis of the carrying capacity of the network, and these computerized studies are becoming essential to permit short-term planning in the low- and medium-pressure networks, medium-term planning for the regulator stations and long-term planning for the high-pressure mains.

## Recent Commercial Considerations

Some consumers have rejected natural gas for central heating and enquiries in Belgium into their reasons for the decision yielded the following results, which varied according to area:

20-50 per cent of replies pointed to explosions and leakage and suggested that natural gas is a dangerous fuel.

30-35 per cent said natural gas was expensive and its use was too costly.

The reputation of natural gas as a dangerous fuel stems mainly from press reports of explosions, not only in Belgium but in Great Britain, France, and Holland, although some of the most dramatic of these were in fact attributable to manufactured gas. To combat this situation, it has been necessary to intensify safety measures (cathodic protection, inspection of equipment, appliances, and installations, leakage detection and repair, etc.) and make this known to the public so as to improve the image of the fuel. The adoption of gas central heating over the coming years should show quite clearly that these criticisms are not objective.

As regards the price, it is quite evident that a rapid penetration of the market by gas can only be achieved by the introduction of tariffs which are genuinely competitive with oil, particularly for central heating. Increasingly competitive oil prices to domestic consumers have put gas distributors in a dilemma:

— Either they can maintain present prices, resulting in a slower penetration of the market and the financial consequences this would have in the medium term.

— Or reduce these prices to ensure market penetration, accompanied by a drop in short-term profits in an already difficult situation.

In view of the recent high cost of conversion, neither solution is acceptable and a compromise must be sought. This could be achieved by determined commercial efforts, together with closer collaboration between the distributors and their suppliers aimed at matching any price concessions which distributors have to make in order to arrive at an adequate development of sales.

Finally, an effort is being made to sell new gas appliances specially designed for Slochteren natural gas in replacement of older appliances. This should result in the elimination of one of the hangovers of conversion, the "awkward" appliance, which was on the borderline of convertibility.

Fig.5 Reception equipment before natural gas

## CONCLUSIONS

In the foregoing paper an attempt has been made to cover systematically the advantages and problems of conversion as they affect the subsequent

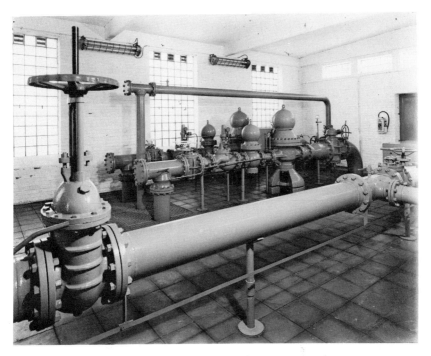

Fig.6 Reception equipment after natural gas

distribution of natural gas, with particular emphasis on our Group's recent experience in Belgium.

The author may be reproached for not putting cost figures to these items but, as indicated earlier, every distribution system has its own individual characteristics. We know from experience that it is a mistake and can be very misleading to apply other people's figures to one's own case, particularly if the application is international, and this is not recommended other than on the basis of detailed comparisons, which are outside the scope of this paper. For this reason, the comments have been limited to a statement of trends and in this final section we shall attempt to draw some conclusions.

### Advantages of Conversion to Natural Gas

These are:
1. It is delivered to the distributor at a lower price per therm than manufactured gas.
2. It is available in large volumes and facilitates the coverage of peak loads.
3. It is normally available at high pressures and makes compression by the distributor unnecessary. At the same time, it favours the development of

medium-pressure networks, which reduces the cost of reinforcements and extensions.

4. The higher calorific value of natural gas nearly doubles the carrying capacity of the network on a thermal basis, which allows a reduction in expenditure on reinforcements.
5. Natural gas is non-toxic, clean, and of constant quality.

On this basis, conversion to natural gas, bringing cheaper gas, reduction in development costs, and an increase in the carrying capacity of the distribution network seems to offer nothing but advantages which can be passed on to the consumer in the form of price reductions for a high-quality fuel, which should, on the face of it, be highly competitive with other fuels.

### Disadvantages of Conversion to Natural Gas

In our operations the reality is less attractive, as the advantages quoted above are not as absolute as they appear to be at first sight.

1. The principal market for natural gas is space heating, which causes significant seasonal variations in sales. Since the price of gas is tied to load factor, it will become progressively more costly both to buy and to distribute, as the utilization factor of the distribution network is also affected.

   Furthermore, to the purchase price of the gas must be added a total charge of $\pm$ 20 per cent of the cost price to amortize the cost of conversion.
2. The increased availability of the gas supply was only made possible by the creation of new supply points. The expenditure necessary for the construction of these reception stations, and the modification of existing reception stations, would probably have been needed in due course in any case, but the arrival of natural gas accelerated it.
3. The increased pressure and dryness of natural gas increased the physical leakage from the network. Unaccounted-for gas expressed as a percentage of sales which, with manufactured gas, was of the order of 7 per cent, rose with natural gas to about 20 per cent. This is unacceptably high from both the financial and safety points of view, and distributors were obliged to condition the gas so as to prevent desiccation of the joints in the mains, and promptly to intensify the detection and repair of leaks, which is costly, and will certainly have to be maintained for some years at a level about three times greater than was necessary with manufactured gas.
4. If it is true that the carrying capacity of the network is nearly doubled, it is nevertheless a fact that a large part of the development in sales is through extensions to the low-pressure network, and through a mainly new medium-pressure network operating at 5 Bar. As a result, the annual rate of investment has not diminished since the introduction of natural gas.
5. In spite of its greater cleanliness and constant quality, the public image of natural gas is not as good as that of manufactured gas.

Fig.7 Equipment rendered obsolescent by natural gas

Because it contains no hydrogen. natural gas gives a "lazy " flame to which housewives find it difficult to adapt, and although not toxic it is considered by the public, unjustly so, to be more dangerous than town gas, mainly because of the wide press coverage of explosions.

Finally, although natural gas is generally considered to be essentially non-polluting, it is increasingly incriminated by local authorities who claim against distributors for causing the deaths of trees. These claims are sometimes justified, but often natural gas distributors are used as a whipping boy for the deaths of trees due to other sources of pollution and to natural causes, which does not improve public relations.

## FOOTNOTE

In the light of the above, it will be seen that the intrinsic advantages of natural gas are to some extent compromised by the consequences of converting the systems from manufactured gas, particularly in the short term. Most, if not all distributors who have converted in the past on both sides of the Atlantic have passed through this stage. In some cases it has been more difficult than in others, depending on local circumstances, but generally the benefits have been progressive rather than immediate.

All gas distributors regard conversion as a painful and expensive malady, but it is self-imposed in the light of complex feasibility studies which each company has had to make when considering its own case, based mainly on the considerations set out in the present paper. In practically every case it is clear that the company passes through a difficult period immediately after conversion, but it is the springboard for a new era in its history.

A distributor who converts to natural gas is jumping in at the deep end. It is an act of faith and one of determination to take advantage of the new supply potential, overcome the very real difficulties involved, and exploit every avenue of potential sales to justify the decision economically.

To do this a dynamic price policy must be evolved, aimed at expansion where gas can be competitive, vigorously and effectively promoted to the consumer, so as to win his attention and direct his thoughts and his plans increasingly in favour of natural gas.

## DISCUSSION

In opening the discussion, *P. F. Corbett* (British Petroleum Co. Ltd) thanked the author and his colleagues for the paper and confirmed their belief that the reason for the invitation to present this paper was to hear the experiences of a fully converted system.

Of the paper, he commented that they had perhaps overemphasized the options and that in reality they had no alternative but to convert.

In 1963 all of Northern Europe was thinking about conversion and in less than nine years ten million customers had been converted. The sums therefore were simple — overnight a five-fold increase was possible from increased pressure and doubling of the calorific value.

Turning to the burning characteristics, he thought the author might enjoy a longer discussion on the "lazy flame" with natural gas and particularly its implications on pilot lights.

When reviewing the overall costs, he pointed out that no credits were apparent for the real estate made redundant by the change-over.

An interesting and perhaps surprising fact given in the paper was that there were no significant increases in sales following conversions.

Before concluding Mr Corbett referred to the relatively short period of only ten weeks allowed for complaints in their conversion programme; this, he considered, was probably inadequate.

In conclusion he expressed the meeting's appreciation of the way in which the author had taken everyone into his confidence and trusted that his firm's "act of faith" (to quote from the paper) would be properly rewarded.

*H. Jagger* (Esso Petroleum Co. Ltd) described his experiences 32 years ago, when he was called upon to convert a number of houses from town gas to bottled gas following a bomb explosion during the war; the work was completed at a fraction of the £38 stated by the author and using only a couple of spanners to adjust the orifices.

He then asked the author why there were no credits in his economics for the increased capacity due to increased CV and allowances for pipe maintenance, replacement, and repair.

*Mr Buckley* envied Mr Jagger his task and pointed out that he had to first train 500 non-skilled and even non-technical people to do his work, which contributed significantly to the costs, particularly as the safety of the appliances had to be guaranteed. Due to the lack of experience of the crews, they had to suspend operations whilst they built up stocks of materials with which they were familiar and modify their plans to accommodate this change.

As to the credit for the pipe runs saved, he thought in fact the additional transitional lines laid probably exceeded any possible savings by way of a capacity bonus.

The "lazy flame" called for a considerable amount of re-design because the alternatives were either too ugly or too expensive. The difficulty was with low flame heights at not full on position and the pilot light flame must be accepted as a fact of life.

Turning to the real estate credits, the rusting gas holders were there for anyone to buy.

*G. Vanderschueren* (S. A. Electrogaz) answered the question concerning public relations and advised that the ten weeks was the allocation of complaint time for the conversion teams; thereafter it became a central function. In fact, they maintained a site office for this initial period; thereafter, complaints were received during the next six months, often only appearing when the winter heating load commenced. They also wrote to customers seeking any complaints.

*Mr Hall* asked how the author distinguished between meter lows and leakages.

*Mr Vanderschueren* advised that the losses were determined by the purchased volume minus sales volume. The effect of variation in temperature and pressure from purchase to point of sale was exactly self-compensating.

In answer to Mr Corbett and Mr Jagger concerning CV credits, Mr Vanderschueren pointed out that, whilst the values were doubled, in theory the sales were halved. He advised that a dcf rate of 10 per cent had been used but that was their 1966 rate and no longer applicable.

*R. Evans* (Gas Council) asked the author to explain the apparent contradiction in the gas holders which had been made redundant, yet his slide showed suspended spheres. This appeared to be inconsistent.

On the question of load factor curve, he said the Gas Council found this interesting, although it did not reflect their expectations.

Referring to leakage, he was surprised to see such a high figure as 5 per cent, increasing to 20 per cent after conversion, as this was not mirrored in the U.K. However, he postulated that there were perhaps fewer variations in equipment and pipeline in the U.K. and, furthermore, there were no asbestos mains.

*Mr Buckley* said the gas holders were low pressure, whereas the high-pressure spheres were a necessary element in any distribution system. He knew of no one who was currently building gas holders.

On load curve, he commented that this reflected the developing and increased usage for central heating and he was hopeful for the future.

# Gas Sales to the Domestic Market

By A. I. D. FRITH

*(The Gas Council)*

## SUMMARY

The paper opens with a brief review of the energy market in Great Britain, with particular reference to the domestic sector. The heating and hot water markets are shown to be most significant for increasing sales of Natural Gas.

Comment on the background of Natural Gas in Great Britain is included and leads into description of the growth pattern and various uses of gas in the period 1960-71. The possible pattern of use in the future is referred to.

The relationship between an energy supply industry, appliance manufacturers, and other relevant enterprises is dealt with on a philosophical rather than a specific basis in view of the complexity of the subject and the space available. The views expressed here are those of the author and are not necessarily shared by colleagues in the Gas Industry.

## INTRODUCTION

Whatever the energy form, the sales potential in the domestic market is governed by historical, economic, social, and climatic conditions.

Natural gas, because of its physical characteristics and constituents, is a more convenient and versatile form of domestic energy than either solid fuels or oils. Assuming adequate availability, penetration of the market therefore depends upon relative value for money and marketing ability. In this context "relative value for money" is taken to mean the cost to the customer of acquiring the equipment necessary to utilize the fuel and its cost in use.

In countries where a distribution system for manufactured gas was already established, the introduction of natural gas has often been followed by a rapid acceleration of domestic sales, usually coincident with an even greater rate of growth in the industrial and commercial sectors. The high calorific value of

165

natural gas compared with manufactured gas, the need to balance sales with supply contracts, and to optimize the load profile to improve the return from capital invested are two of the more significant reasons for this. With the massive increases in sales that follow the introduction of natural gas, it is essential that the right level of sales is reached in each market to achieve the optimum mix.

It is against this background and by drawing examples from the British domestic energy market that this paper is written.

## ENERGY IN DWELLINGS

Gas is but one of four sources of energy available to many dwellings. Of these, electricity is the only one capable of meeting all the requirements in a modern home, directly in the form in which it is distributed. The marginal cost of additional electricity is high and rising because of the low efficiency of fossil fuel conversion and the high capital cost of nuclear production. It would appear then that availability and price, which have limited the growth in sales in the past, will continue to do so for many years to come. Gas, solid fuels, and oils will therefore continue to meet the demands of the domestic market but in changing proportions.

Before considering the part which gas can play, it is as well to look at the domestic market for energy as a whole.

### Availability

It is assumed that, of the four energy forms, solid fuels and oils will be available to all households. In Britain this is not quite true for electricity and still less so for gas, as the figures in Table I demonstrate. These figures are included as background to later remarks.

### TABLE I
**Gas and Electric Supplies to Households in 1971**
(Sources: "Audits of Great Britain"
Surveys and Gas Council Statistics)

| | |
|---|---|
| Total number of households | 18.1 million |
| No. of households connected to electricity | 17.9 million |
| No. of households connected to gas (*i.e.* using gas meters) | 12.8 million |
| No. of households in gas supply area* | 15.6 million |

* Defined as households where a gas main is available to the road in which the household is situated.

### Usage

It is also helpful to review the total domestic usage of energy, both as it is at the moment and as it may develop in the future.

Table II indicates the total consumption of energy in dwellings during the year 1970 for each energy form. It will be seen that the total consumption was 14680 x 10[6] therms. For comparative purposes it should be noted that during the same year the total consumption of energy, in all forms but excluding transport, was 6676 x 10[6] therms in the commercial market and 25421 x 10[6] therms in the industrial (final user basis). The Ministry statistics in Table II are on the "heat supplied basis" and include Northern Ireland, which forms a very small proportion of energy consumption (less than 1 per cent in the case of gas).

### TABLE II
### Energy Consumption in Dwellings
(Source: Ministry of Technology, Digest of Statistics, 1971)

| Energy form used | Therms supplied | % of total |
|---|---|---|
| Solid fuel | 7137 x 10[6] | 48.6 |
| Gas | 3583 x 10[6] | 24.4 |
| Electricity | 2625 x 10[6] | 17.9 |
| Oil | 1335 x 10[6] | 9.1 |
| Total energy consumption in dwellings | 14,680 x 10[6] | 100 |

The growth of the total market for energy in dwellings derives from new homes built plus improvements to the standards of comfort, convenience, and leisure in existing dwellings. Against this must be offset energy usage lost from dwellings demolished and the increased efficiency of modern appliances in the utilization of energy. For example, in Great Britain in 1965, 60 per cent of the home's energy was supplied by solid fuel and 13 per cent by gas. Five years later, solid fuel was supplying only 48 per cent and gas had increased its share of total energy usage to 25 per cent.

This movement towards the use of more efficient energy forms has meant that the total growth of energy usage in the domestic market in Great Britain has increased by only 1.6 per cent net therms during the five years, 1965 to 1970. However, whilst conversion of electricity to heat, light, or power is generally a highly efficient operation the increased use of electricity in the home does not necessarily lead to an overall reduction in primary fuel usage. The low efficiency of fossil fuel conversion to electricity at the power station is the reason for this.

For a quality, convenience fuel, such as natural gas, the best opportunity for increasing sales quickly in a slowly growing but large existing market is to win business by replacing less desirable energy forms. Table II shows the form in which energy was supplied in 1970 and clearly it is not necessary to debate here the vulnerability of solid fuel. It is important, however, to recognize that energy is used in dwellings in a variety of ways and to quantify each of these uses as a step towards developing a viable marketing strategy and tactical plan for natural gas.

The way in which energy is used will vary considerably from one home to another. Nevertheless, it is possible to estimate the breakdown of energy usage in a home which is provided with full heating, hot water, cooking, lighting, and electrical power services and that these facilities are used to achieve modern standards of comfort. This is shown in Table III.

### TABLE III
### Breakdown of Energy Usage in a Dwelling
(Source: Gas Council Estimates)

| Energy usage | Therms per year | % of total |
|---|---|---|
| Heating | 770 | 70 |
| Hot water | 210 | 19 |
| Cooking | 70 | 6½ |
| Electric lighting and small power | 50 | 4½ |
| Total | 1100 | 100 |

### Potential

The overwhelming importance of the heating and hot water markets to the fuel industries is illustrated by Table III. Further analysis of the market is also

### TABLE IV
### Establishment of Central Heating Systems (December 1970)
(Source: Audits of Great Britain Surveys)

| | |
|---|---|
| Total number of households | — 18.1 million |
| Total number of households with central heating | —  5.8 million |

| Fuels used | % of total |
|---|---|
| Gas | 33* |
| Solid fuel | 30* |
| Electricity | 26 |
| Oil | 8 |
| Communal | 3 |

*Of sales made between October 1970 and September 1971, gas had a 44 per cent share and solid fuel 10 per cent.

illuminating. For example, the establishment of central heating systems and main living room fires at the end of 1970, as shown in Tables IV and V, indicates potential market opportunities more specifically.

Whilst markets other than heating and hot water are also important, they do not offer anything like comparable load growth opportunities. For example, with some 11 million gas cookers in use from 12.8 million homes connected to gas, the opportunity for further penetration is limited. Nevertheless, retention of the revenue from cooker usage and the importance of the cooker as a springboard for additional appliance sales make this a very important appliance group. Recent research has proved that an existing gas customer is much more likely to buy additional gas appliances than a householder not currently using gas.

## TABLE V

### Establishment of Main Living-Room Heating* (December 1970)
(Source: Gas Council Estimates)

| Fuel used | No. of homes | % of total |
|-----------|-------------|------------|
| Solid fuel | 9,100,000 | 51 |
| Gas | 5,000,000 | 28 |
| Electricity | 3,400,000 | 19 |
| Oil | 250,000 | 2 |

* Where central heating is not installed or where a separate heater is installed in the main living-room to supplement central heating.

## GAS IN GREAT BRITAIN

**Natural Gas**

Small quantities of seaborne natural gas have been imported into Great Britain from Algeria since October 1964, but it was only in October 1965 that substantial indigenous reserves were discovered under the North Sea. Before the end of 1970 a national high-pressure transmission system had been constructed and was conveying natural gas directly from the sea terminals to each of the 12 Area Gas Boards. The Boards in total sell gas to 13.4 million customers throughout England, Scotland and Wales.

By April, 1971, natural gas accounted for 80 per cent of all gas available for supply by the nationalized British gas industry and, of total sales of gas in the year ending March 1971, 32 per cent was in the form of direct natural gas supply. The conversion of customers' premises to direct supply of natural gas is proceeding at the rate of 2 million/year and will be virtually completed by the end of 1975.

During the five years prior to the discovery of North Sea gas, the industry completed its transition from manufacturing gas from coal to the lower cost of manufacturing gas from oil feedstocks. As this transition had progressed, so had the industry's ability to compete more effectively in the market place. The discovery of natural gas therefore came when the industry was already in a position of strength.

## Sales of Gas

Domestic, commercial, and industrial gas sales during the period 1969-71 are shown in Appendix 1. It will be seen that during this period the bulk of the industry's market and growth was in the domestic sector. Because of the introduction of natural gas, this pattern is now changing rapidly.

## Pattern of Domestic Gas Usage

The breakdown of 1970-71 domestic gas usage and the probable pattern of usage by 1975-76 is shown in Appendix 2. This again illustrates the significance of the heating and hot water load.

## Gas Pricing

As mentioned earlier, it is necessary to evolve an overall marketing strategy which will embrace the total operation of the enterprise in relation to the single product, in this case gas. Within the total marketing plan will be embraced the pricing strategy. The relationship between the mix of loads and the mix of prices is critical if financial as well as load growth objectives are to be achieved.

Competitive energy prices in the various markets will, if considered against the "premium" values of the various forms of energy, provide the parameters within which the gas prices must be set to be consonant with specified load growth objectives. Price elasticity studies (assessment of change in volume of sales at different price levels) are a fundamental step in preparing a marketing strategy for gas as with any other competitive product.

In Britain gas is sold to domestic and smaller commercial and industrial customers at published tariff prices. For customers using more than 100,000 therms/year, special contracts can be negotiated. Contracts are based upon a market-related pricing philosophy, with the proviso that no contract is ever negotiated if it is impracticable to achieve a realistic margin above cost base.

In supplying gas to the domestic market, the costs can be broadly stated as falling into three areas, namely, customer, capacity, and commodity costs. Tariff structure is promotionally-based, in as much as the customer benefits from a lower price per therm when he increases his usage. This lower price reflects the reduced system unit costs and other overheads associated with increased usage.

It has been demonstrated that heating and hot water are the areas offering the greatest opportunity for growth and, typically, gas is sold for these purposes on a two-part tariff. The customer's costs and a proportion of the capacity costs are recovered within a standing charge and the remaining charges are reflected in a commodity rate which will be considerably below the published "flat" rate for small users. "Budget billing" is another promotional aid, which permits a customer to pay a standard monthly or quarterly sum against his estimated annual consumption.

Bearing in mind that the pattern of load is now changing rapidly, it is still of interest to note in Table VI the total income and average income per therm achieved from sales of gas to the domestic, commercial, and industrial markets in Britain during the financial year 1970-71.

TABLE VI
**Income from Sales of Gas (1970-71)**
(Source: Gas Council Annual Report and Accounts 1970-71)

| Market | Total gas sales ('000 therms) | Total income (£'000) | Average income per therm (p) |
|---|---|---|---|
| Domestic | 3,653,098 | 386,329 | 10.58 |
| Commercial | 775,583 | 69,912 | 9.01 |
| Industrial | 1,704,235 | 77,049 | 4.52 |

### Safety of Natural Gas

The word "gas" is itself emotive. Accidents, a number of which are inevitable with the widespread use of any form of energy, are therefore more likely to be newsworthy with gas than with other energy forms. The conversion of manufactured gas appliances to use natural gas is also of considerable news value, since it represents an imposed requirement on all existing customers.

As a result of public unrest about the safety of natural gas, a government inquiry was held in 1970. Professor Morton was appointed by the Minister of Technology to undertake the inquiry and his findings were published by HMSO as "Report of the Inquiry into the Safety of Natural Gas as a Fuel".

This report, which is comprehensive and authoritative, is recommended reading for anyone interested in this topic. The conclusions in the report are reassuring; the first states that "Natural gas can be stored, distributed, and used with safety in correctly designed and properly maintained equipment". On the subject of conversion, the Report concludes that "the conversion operation has been carefully planned, well organized, and competently executed. The labour force has been adequate in numbers and quality".

The main criticism raised by Professor Morton concerned the "lack of adequate ventilation and flueing", though he conceded that progress was already being made in this area and that "it can be expected that the customer's premises are left in a safer condition after conversion than before".

Accidental deaths in England and Wales attributed to carbon monoxide poisoning from gas have fallen from 1327 in 1963 to 748 in 1967 and to a provisional figure of 320 for 1970. This last figure represents 1.7 per cent of all accidental deaths in 1970.

Despite these encouraging statistics, it should be recognized that natural gas, although non-toxic, is capable of causing serious explosions in certain circumstances. It is important that this should be understood, if only to prevent unfortunate accidents in association with attempted suicides. There are a number of cases on record where innocent parties have died in explosions resulting from suicide attempts.

From international experience it can be positively stated that, properly "engineered" and used, natural gas is a safe fuel, relative to other energy forms and usage.

## MARKETING STRATEGY

In Great Britain the prospect of replacing manufactured gas with natural gas, from virtually a single source, determined the need for planning on a national basis. To this end, the Gas Council formed Marketing, Production and Supply, and Economic Planning Divisions during 1968. From that time the 12 Area Boards, which had hitherto been working independently within broad national guidelines, have closely co-operated with the Gas Council in the development and implementation of an integrated strategy and in corporate planning.

From factors, some of which have already been discussed in this paper, a marketing strategy was developed to make the best use of the large but finite supplies of gas then available. For an extractive industry with the probability of further discoveries and thereby additional availability of natural gas, the overall strategy by necessity had to be flexible. In the event, the strategies adopted by the industry in 1968 can be stated broadly as follows:

—To sell as much gas as possible to domestic and small commercial and industrial customers at published tariff prices. To sell the remainder up to the planned total to large customers ($a$) on individually negotiated "firm" supply contracts at market related prices and ($b$) to optimize the load pattern by selling "interruptible" gas at special contract prices.—

For the domestic market, analysis of present and potential energy usage, competitive activities, and likely developments provide the foundations on which the domestic marketing strategy was built. From the data already put forward, the potential markets for central heating and room heating stand out. With almost 11 million gas cookers using relatively high-priced gas, it is also important to retain this load which, by 1975, will still represent about 16 per cent of the total domestic load. Thus a brief statement of the domestic marketing strategy can be given as follows:

—To maximize sales of central heating and gas fires whilst retaining the cooker load and developing other markets for domestic gas appliances.

## PLANNING FOR PENETRATION OF THE DOMESTIC MARKET

So far, this paper has been concerned mainly with consideration of the domestic energy market as a whole, the place of gas within it, and the determination of viable broad strategies. Applying similar methods of assessment to each sector of the domestic market highlights opportunities for sales of gas and appliances, product developments, etc. In turn, this enables tactical planning to be carried out and to be implemented.

### A Planning Framework

The planning framework which the Gas Council has used includes detailed analysis of the market place under the following headings.

### Property

| | |
|---|---|
| New homes | — Number and type built for private owner occupation. |
| New homes | — Number and type built for local authority renting. |
| New homes | — Number and type private and local authority with communal heating/hot water schemes. |
| Existing homes | — Number and type owner occupied. |
| Existing homes | — Number and type local authority owned and rented. |
| Existing homes | — Number and type privately rented. |

### Replacement

Replacement of appliances fired by other fuels and also of outdated gas appliances is another aspect of market planning which is given special attention. Up-to-date figures are therefore maintained for appliance establishment and sales both for gas and its competitors. When it becomes apparent from these records that a market is becoming saturated, the new homes market stands out as the principal remaining growth area.

### Social Classification

The type of appliance or heating system that a customer can afford will greatly depend upon his income. A socio-economic classification within each property group is therefore maintained.

### Other Factors

Age of property and the age of the housewife are some of the other factors considered.

## GAS APPLIANCES

The attitude of domestic customers towards gas is not only governed by the qualities of the fuel itself, but also by the service it can provide in association with an appliance. This symbiotic relationship between the fuel producer and appliance manufacturer means that the success of each is very dependent on the success of the other.

Clearly, it is necessary for the purveyors of energy to liaise closely with appliance manufacturers, wholesalers, retailers, installers, and after-sales service organizations.

### Liaison with Appliance Manufacturers

Whether nationalized or not, the energy purveyors are in business. By definition, therefore, one of their main objectives must be profit, since this is the principal measure of success in a competitive economy. This is also true of appliance manufacturers and other associated interests, but sometimes the legal and social constraints differ.

It should be recognized that, although interdependent to a degree, the more specific objectives of fuel purveyors and appliance manufacturers may not always be in harmony. For example, the gas industry has for several years devoted a higher proportion of its marketing resources to penetrating the heating market than was the case hitherto. Manufacturers of other than heating appliances who did not compensate for this may have suffered. The point is that as the objectives of individual but associated enterprises change, the market balance is inevitably influenced: politically-inspired change can also lead to distortions. Whatever the cause, significant changes in policy must be identified and their effects anticipated or someone somewhere suffers in the pocket.

Establishing a workable balance of resource expenditure among interdependent enterprises has and always will be a difficult task. It is almost inevitable that the various enterprises will criticize each other's policies, especially when business is not flourishing. However, it does not necessarily follow that perfect harmony is a preferable alternative: a certain amount of competitive friction may in the medium and long term be beneficial to the customer.

Development of new appliances, particularly "breakthrough" appliances, may be very much in the interests of the energy purveyor. It may not, however, be in the immediate interest of the appliance manufacturer who has just tooled up at considerable expense to produce a new appliance of similar purpose but of traditional design and/or performance.

Due to take-overs and mergers, larger manufacturing companies are emerging who, by way of product diversification, make appliances for several or all energy forms. Is it in their interest to perpetuate the manufacture of minority lines? If not, what influence is this likely to have upon the market situation of the fuel interests?

The essence of marketing is the realistic assessment of potential for any product or product group in a defined market; assessment of the various resources needed to achieve targeted penetration; preparation of a workable plan and action programme; implementation of the programme and the setting-up of a control system to monitor results and make necessary adjustments.

Short of complete vertical integration for a "self-contained" product by a manufacturer, consultation between all groups immediately involved with a particular product group is the only practical step which can lead to an understanding of the marketing situation by all concerned.

In a free enterprise society competition can and does result in customer benefits. On the other hand, completely unfettered competition can result in low standards and create social problems. It can also be wasteful of national resources.

Within the context of a single nation, competition and social conscience eventually result in the emergence of a limited number of responsible large enterprises each in competition with the other; all in competition with related products and for a share of the customer's total purse.

Internationally, evolution within "like" societies follows a similar pattern, until eventually a limited number of international companies control a high proportion of the total world market for their product groups. This has happened with the motor car and seems likely to happen with relatively high price durable products such as fuel appliances.

This may seem a far cry from selling gas and gas appliances, but at the end of the day the climate in which sales need to be achieved is determined by governments — as we have recently seen in relation to the nationalized British gas industry.

In general, the larger the enterprise and the nearer it comes to complete vertical integration, either directly or by franchise, the stronger will be its competitive position. The British gas industry — probably the most integrated of its type in the Western World — during the past four years has been able to exercise a very strong influence over the totality of domestic gas and gas appliance marketing. The industry weathered the storm of competition when it was wedded to coal in the 1940s and early 1950s; innovated itself out of trouble in the late 1950s and 1960s until today it is the biggest in the Western World outside the U.S.A. Its size may be of special interest to European appliance manufacturers and others, with obvious implications as Britain integrates with the European Economic Community. Appendix 3 gives some comparisons of the gas industries in Great Britain, France, Western Germany, Holland and Belgium.

## CONCLUSIONS

Natural gas is a highly desirable and safe form of energy. Its physical characteristics and chemical constituents make it especially suitable for domestic and other purposes requiring fine control and automatic operation.

It is emphasized that targets for domestic, commercial, and industrial sales of natural gas must be balanced in the right "mix" to achieve an optimum return on capital resources as well as to provide a wider range of markets.

Given adequate availability of gas and gas appliances at reasonable prices, proficient marketing will result in further substantial penetration of the energy market. In the domestic sector, growth will result mainly from heating and hot water sales to new and existing dwellings. In the latter, natural gas will often replace solid fuel.

Natural gas, because it has a higher calorific value than manufactured gas, enables existing pipework systems to carry more energy with only very limited additional capital expenditure. The marginal contribution from additional sales is one reason why in countries such as Great Britain, where there is already a well-developed manufactured gas industry, continued effort will be applied to retain the use of gas for cooking and refrigeration and to introduce new appliances. One such new product group is leisure appliances, including greenhouse heaters, patio lights and barbecues, swimming pool heaters, and garden incinerators.

Sales of natural gas to the domestic market are dependent upon sales of appliances, their installation, and after-sales service. The need for co-operation between the various enterprises concerned is apparent and consultation is an important component of successful marketing if the needs of the customer and the enterprises are to be met.

Further rationalization of enterprises within a European framework is likely, especially in appliance manufacturing and possibly in merchandising through retail outlets.

## ACKNOWLEDGEMENT

The author would like to thank the Gas Council for permission to publish this paper and his colleagues who have assisted in its preparation.

## APPENDIX I
### Gas Sales in Great Britain by Domestic, Commercial, and Industrial Categories
### 1960-61 – 1970-71
#### (million therms)

| | 1960-61 | 1961-62 | 1962-63 | 1963-64 | 1964-65 | 1965-66 | 1966-67 | 1967-68 | 1968-69 | 1969-70 | 1970-71 |
|---|---|---|---|---|---|---|---|---|---|---|---|
| Domestic | 1 291.2 | 1 345.5 | 1 493.3 | 1 553.7 | 1 726.8 | 2 005.5 | 2 267.2 | 2 652.2 | 3 010.8 | 3 362.1 | 3 653.1 |
| Commercial | 469.2 | 480.8 | 521.6 | 508.9 | 527.3 | 550.6 | 579.0 | 632.5 | 676.5 | 714.2 | 775.5 |
| Industrial | 851.8 | 856.8 | 852.2 | 861.2 | 915.3 | 928.2 | 908.4 | 914.5 | 976.4 | 1 159.1 | 1 704.2 |
| Total | 2 612.2 | 2 683.1 | 2 867.1 | 2 923.8 | 3 169.4 | 3 484.3 | 3 754.6 | 4 199.2 | 4 663.7 | 5 235.4 | 6 132.8 |

Source: The Gas Council Annual Report and Accounts for the financial years 1960-61 to 1970-71

## APPENDIX 2
### Breakdown of 1970/71 Domestic Gas Load in Great Britain by Usage and Probable Usage Pattern by 1975/76

1970/1971

TOTAL
3655.1 x 10⁶ therms

| CENTRAL HEATING PLUS COMBINED WATER HEATING 37.2% |
| GAS FIRES 26.5% |
| WATER HEATING 8.1% |
| GAS COOKERS 25.2% |
| OTHER 1.2% |

1975/1976

| CENTRAL HEATING PLUS COMBINED WATER HEATING 51.5% |
| GAS FIRES 25.2% |
| WATER HEATING 6.5% |
| GAS COOKERS 16.0% |
| OTHER 0.8% |

Source: Gas Council Estimates for the financial years 1970/71 and 1975/76.

## APPENDIX 3

### Some Comparisons of the Gas Industries in Great Britain, France, Western Germany, Holland and Belgium

| Country | Year | Population in Gas Supply Area | | Yearly Aver. Consumption Per Domestic User (therms) | Appliance Establishment (excluding) (000's) (bottled gas) | | |
| --- | --- | --- | --- | --- | --- | --- | --- |
| | | (000's) | Of total population % | | Cookers | Refrigerators | Heating† Appliances |
| Great Britain | 1960 | 45,000 | 88 | 106 | 11 000 | 250 | 3 100 |
| | 1969 | 47,300 | 84 | 234 | 10 900 | 1 100 | 9 650 |
| | 1971* | N.A. | N.A. | 317 | 10 800 | N.A. | 11 487 |
| France | 1960 | 25,430 | 56 | 73 | 6 010 | 30 | 783 |
| | 1969 | 31,961 | 63 | 140 | 7 220 | 9 | 1 876 |
| Western Germany | 1960 | 33,940 | 61 | 62 | 7 260 | — | 370 |
| | 1969 | 37,921 | 62 | 156 | 7 000 | — | 2 065 |
| Holland | 1960 | 9,531 | 82 | 92 | N.A. | N.A. | N.A. |
| | 1969 | 12,483 | 96 | 491 | N.A. | N.A. | N.A. |
| Belgium | 1960 | 5,820 | 63 | 62 | 1 500 | 30 | 179 |
| | 1969 | 6,298 | 65 | 150 | 1 579 | 64 | 760 |

Source: Mainly from Domestic Gas Marketing Statistics, 1969. Report by the French Delegation of the International Colloquium about Gas Marketing (1971)

N.A. = Not Available
* 1971 figures only available for Great Britain
† Total of figures for gas self-contained separate appliances and gas central heating appliances.

## DISCUSSION

*F. De Camps* (Calor Gas Ltd) expressed his appreciation of the very great honour it was for him to open the discussion on the excellent paper by Mr Frith.

He explained that he had been purveying a gas (LPG) for 25 years, yet its only mention was by way of an exclusion at the end of the ultimate appendix.

He pointed out that under "Availability" the author stated that gas was not available to every household in Britain; this was not true, as bottled gas was available.

Table I showed that 2·5 million households were denied the option of piped gas in 1971 and Appendix 3 showed that the population in the gas supply areas fell from 88 per cent in 1960 to 84 per cent in 1969. Mr De Camps felt that this was an area in which the free enterprise LPG business should be encouraged to work more closely with the nationalized gas industry in an effort to give home business a full and free option.

He went on to cite the examples of Wick and Thurso, where his company had replaced the conventional gas industry system with a butane air plant and turned a £30,000/year loss into a modest profit as agents for the Scottish Gas Board.

Turning to marketing strategy, he saw that some Gas Boards, who had hitherto been playing down the cooker loads, were now modifying their policies. He stressed how important this was for cooker manufacturers in the U.K. whom he believed had had a very difficult time since the advent of natural gas, as their major role had so far been the depressing one of manufacturing, at the lowest possible prices, conversion parts for all of their cookers in the field. These parts, which were supplied free of charge to the consumer, represented the main working parts of a cooker and, in consequence, must have played havoc with the replacement cooker sales. These sales were, it seemed, picking up, according to the latest figures, and he expressed the view that he hoped this trend would continue because it formed a prerequisite for a firmly based export business which, with the Common Market coming, must be in the national interest.

The paper, he pointed out, said the sales potential in the domestic market was governed by historical, economic, social, and climatic conditions and he thought that for either the British appliance manufacturers to sell in Europe or for the Europeans to sell in Britain there would have to be considerable attitude research. For either side to be successful in, for example, the cooker market, then either the British housewife must accept the grill in the oven or the European must accept the British grill.

He referred also to the water heater, which in large areas on the Continent took the place of our traditional gas-fired boiler and to break into this market was going to call for the highest degree of professional marketing from the appliance manufacturers.

The section dealing with the liaison with appliance manufacturers was, he thought, one of the most important in the paper which, to quote, was "The essence of marketing".

In his experience in many Board areas he could not recall one attitude study or one questionnaire which asked the housewife if she preferred one or two simmering burners, or should they be at the back or front of the hot plate, etc.

He believed that with its unique contact with the consumer, this was the task of the gas industry, who should in turn, like a good salesman, relay back to the manufacturing industry the requirements of the customer and then give wholehearted support to those manufacturers who set out to meet the consumer's needs.

He concluded that the nationalized gas industry in Great Britain had a record of service to the public which was second to none and that the oil industry, with its tremendous resources, its courage, and its foresight, had provided them with the highest possible quality gas.

The British gas appliance manufacturer still gave the best value for money.

He considered that for complete success a professional marketing outlook by the gas industry was needed and that this paper was a forerunner of such a policy.

*M. W. H. Peebles* (Shell International Gas Ltd) commented that the central heating load was a vital part of any energy selling business and with the present gas tariffs they were cheaper than electricity, coal, or oil for the majority of the market.

He asked, therefore, with 1975 just round the corner, what the future held for central heating.

*Mr Frith* considered that the figures quoted in the papers earlier that day were about right, although perhaps a little optimistic.

*J. M. C. Bishop* (Phillips Petroleum) referred to the apparently poor PR aspects of the gas industry, quoting such examples as unpresented bills. He asked how this was being developed to improve the gas industry's public image.

*Mr Frith* replied that they were well aware of this. They had made mistakes in the past and this had attracted often unwarranted press coverage. He assured the meeting that this was under constant review.

*Miss M. P. Doyle* (Esso Petroleum Co. Ltd) asked what percentage of the industrial sales would be interruptible in 1975.

*Mr Frith* replied that he was not prepared to answer this, except to say that it would be greater than 20 per cent.

*P. F. Corbett* (British Petroleum Co. Ltd) asked if it was realistic to consider appliances made for the U.K. market, where the Cu was 1000 Btu/cu ft, would be suitable for areas served by the Dutch gas having a Cu of 870 Btu/cu ft.

*Mr Frith* replied that there would be no problems in this respect, as the equipment was basically interchangeable.

Mr Frith, in summarizing the comments and replying to Mr De Camps, agreed that LPG had completely eluded him and it was a serious omission. All the more so as six Area Boards were in the bottled gas business.

There was an omission under "pricing" in that no mention was made of market related prices and the interaction of contracts. There was a balance to be found as the sales mix changed to opt out of the low price contracts and concentrate on the premium markets.

He agreed that they had soft-pedalled a little on cookers, but with their 1000 showrooms throughout the country their sales of all hardware were important. However, it should be noted that 25 per cent of all cooker sales were sales associated with conversion.

The market research aspects were and are considered and explored in considerable depth and made known to the manufacturing industry. In fact, all new equipment was, after development, given extensive marketing trials.

Referring to the domestic market, he stated that 75 per cent of all homes had less than 1100 sq ft floor area and that gas was therefore the obvious and cheapest form of central heating. There was no room for the oil tank under such conditions. However, for the four-bedroom house and upwards, perhaps oil was the cheapest solution.

He concluded by referring to the asides from Mr De Camps about barbecues and patio lights and stated that they were in fact offering packages and this was significant, since the supply covered all of the various outlets from the same connection.

# Potential for Natural Gas as Fuel or Feedstock in Industry

## By C. HIJSZELER

### *(N.V. Nederlandse Gasunie)*

### SUMMARY

The potential of natural gas as a fuel or as a feedstock in industry is determined by the availability and the supply conditions of natural gas, its competitive position as compared with other fuels, and the development of the energy market.

N.V. Nederlandse Gasunie has been responsible for the transport and the sales of natural gas in Holland since 1963.

A sales plan has been made, based on the conditions prevailing in the Netherlands. The results of the realization of this sales plan are discussed.

In industry, natural gas is used mainly as an under-boiler fuel. If certain technical developments are utilized, natural gas has an appreciable premium in this type of utilization.

There are premium uses of gas in industry; some of these, such as the use in total energy systems, incineration, firing of ceramic products, and direct drying processes, are discussed. Factors that influence the penetration of natural gas in the premium market are, among other things, the importance of the energy budget in the total production costs of the industry, possibilities of product improvement or increase of production rate, pay-out time of the special gas-fired installation, and the opinion of the industry concerning the reliability of these installations.

Though dry natural gas is a good feedstock for a limited amount of chemical processes, such as ammonia and methanol, optimum results for the industry and the gas supplier can be obtained only if certain conditions concerning gas quality and plant location are satisfied.

The profit on the use of natural gas is decisively influenced by technological developments in plant design, in transport of chemical products, and by economical and political conditions in other areas where gas is available.

## INTRODUCTION

In 1959 the first indications were found in Groningen that under the Dutch soil there lay a wealth of energy of, by Dutch standards, unprecedented dimensions. Today we know with reasonable certainty that the total amount available in the Groningen field is about 1800 milliard cu. m. Together with more recent discoveries in the provinces of Drenthe, Friesland, and North Holland, the total amount of gas available can be estimated at about 2000 milliard cu. m. In 1963 the N.V. Nederlandse Gasunie was founded, its shareholders being DSM, Shell, Esso, and the State, and was given the task of transporting and selling Groningen gas in the Netherlands and abroad.

## DEVELOPMENT OF GAS SALES

The development of the sales of Dutch gas has been determined by the principle that optimum proceeds for the Dutch economy have to be obtained.

In order to find the conditions that have to be fulfilled to materialize this optimum, presented by the maximum present-day value of the net proceeds from the existing reserves, models have been used. In these models allowance is made for production possibilities, production costs, and the influence of price variations on the quantities sold. Another factor that has to be taken into account is that customers have to make investments in order to handle the gas. The companies supplying the public distribution market require a guarantee of gas delivery for 25-30 years, in order to be able to afford the investments required in distribution grids. Industrial firms that install new gas-fired equipment or convert existing firing equipment require continuity in gas delivery for 15-20 years to obtain reasonable benefits.

This means that if natural gas has not to be regarded as a very temporary and thereby unattractive source of energy, the gas sales must be continued for about 25 years.

The above studies and considerations led to the conclusion that the largest possible share must be obtained for natural gas in national energy consumption and, as far as the reserves allow, the remainder must be sold abroad. Futhermore, the rate of penetration had to be high, which determines the commercial and technical measures to be taken to realize the sales.

Gas has penetrated rapidly in the Dutch energy market, as Fig. 1 indicates, and will do so in the future. In 1975 it will cover 40 per cent. if the total requirements of primary energy in 1975, which means 60 per cent. of that part of the energy consumption that can technically be substituted by gas.

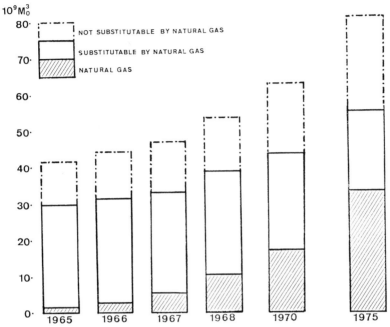

Fig. 1 Natural gas in the Dutch energy economy

## MARKETS FOR GAS

It will be clear that a deep penetration in the energy market, as indicated above, can only be obtained in an industrialized country like the Netherlands if, besides a deep penetration in the domestic energy market, gas is sold in appreciable amounts to industry.

In Fig. 2 the gas sales to different market sectors as realized and expected in the period 1965 - 75 are given. In this figure are indicated the sales to public distribution companies, to industry, and to power stations.

Though the subject of this paper concerns the use of gas in industry some remarks have to be made about the domestic market for better understanding of certain aspects of industrial gas marketing.

The domestic market can bear high gas prices because gas has distinct technical advantages over some competitive forms of energy and/or the prices of the competing forms of energy are high.

Therefore, the domestic market, where gas has a high value, is interesting for a gas sales organization. On the other hand, if the sale of gas in the domestic market is for a major part for heating, it shows an unfavourable characteristic - a sales volume strongly depending on the seasons of the year.

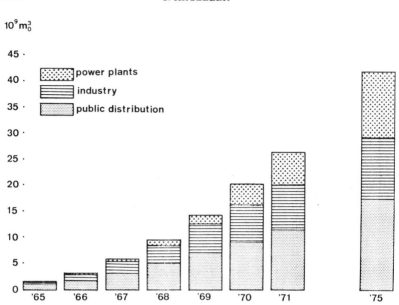

Fig. 2 Gas sales in the Netherlands

This gives rise to relatively high investments in the transport and distribution
grids in relation to the total annual gas sales, as compared with a gas transport
system, which carries a constant load.

The above situation can be improved by sales of gas to markets that
have a more constant need for gas. By doing so, the base load of the grid is
increased and thereby the load factor; a better use is made of the investments in the transport grid.

Further improvements can be made by selling gas on an
'interruptible' basis. In this case the gas is offered to the customer at a
lower price than for a firm supply. The gas supply company, however, has in
this case the right to discontinue the gas supply to the customer in case of
high rate of gas sales to be the domestic market in wintertime.

Other methods to improve the load factor on a transport grid exist,
but are regarded as being outside the scope of this paper.

## THE USE OF NATURAL GAS IN INDUSTRY

For a good understanding of the potential of natural gas in industry,
some characteristics of the industry have to be known. This is illustrated with
the help of some aspects concerning the Dutch industry.

The figures given below are obtained from surveys made by Provincial
Economic and Technological Institutes on behalf of N.V. Nederlandse

Gasunie. Industrial firms can be classified according to size, using their energy consumption as a yardstick.

Fig. 3 illustrates the situation in 1964. Some large chemical industries and other heavy industries are not incorporated. The graph shows that about 92.5 per cent. of all industrial firms had an annual energy consumption equivalent to less than 1 million cu. m natural gas/year and together represented only 17 per cent. of the total consumption, whilst less than 1 per cent. of all industrial firms had an annual consumption of one million cu. m or more and that these firms together accounted for more than 50 per cent. of the total energy consumption.

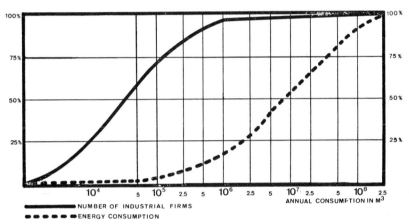

Fig. 3 Industrial firms and their energy consumption classified according to size in 1964 (cumulative)

The picture can be supplemented by the data concerning the type of fuel used to satisfy industry's energy requirements.

The pattern of the total fuel consumption in 1964 is given in Table I. Liquid fuels played a dominant role in the fuel market. Heavy fuel oil had a share of more than 60 per cent. in all industries using more energy than the equivalent of one million cu. m/year. The total energy consumption of these industries represents 70 per cent. of industry's needs.

### TABLE I

#### Industrial Fuel Consumption 1964

| | |
|---|---|
| Solid fuels | 12.9 % |
| Liquid fuels | 79.8 % |
| Gaseous fuels | 5.4 % |
| Electricity for heating | 1.9 % |

A survey of the industry can be completed by data about the types of installations that consume the fuel.

Table II gives a survey of the types of installations used in industry and their importance as fuel users. The important observation to make at this point is that about 70 per cent. of the fuel used in industry is used as under-boiler fuel.

TABLE II

**Percentage of Total Energy Consumption in Industry per Type of Installation and per Industry**

| Industry | Type of installation | | | |
|---|---|---|---|---|
| | Boilers | Kilns | Central heating | Other |
| Food, beverages, tobacco | 87.8 | 7.2 | 2.3 | 2.7 |
| Clothing and cleaning | 86.2 | 1.6 | 10.7 | 1.5 |
| Textiles | 93.8 | 0.5 | 4.6 | 1.1 |
| Leather, rubber | 88.4 | 2.1 | 8.5 | 1.0 |
| Paper | 97.3 | 1.0 | 1.4 | 0.3 |
| Glas, cement, bricks and ceramics | 14.8 | 73.9 | 2.5 | 8.8 |
| Chemical | 80.4 | 13.2 | 2.0 | 4.4 |
| Metal | 28.1 | 39.2 | 25.6 | 7.1 |

For the situation in Holland, the structure of the industry given and a rapid and deep penetration of natural gas in the industrial energy market set as a target, the gas sales had to be concentrated primarily on the replacement of heavy fuel oil used in boilers by natural gas.

It should be realized, that if natural gas is used as an under-boiler fuel, the good qualities of natural gas as a fuel do not show to full advantage.

Apart from this aspect, the price of the gas has to be so low that competition with heavy fuel oil is possible and certainly lower than it could be if its good qualities were used in several other applications.

If natural gas is not available in ample quantity or in case the gas market has already been developed to a certain extent, a situation that approaches in the Netherlands, priority should be given to applications where the economic advantages of gas as a quality fuel are brought out, generally labelled as premium uses. These uses will be discussed in a separate chapter.

In the Netherlands the commercial approach of gas sales was adjusted to the fact that heavy fuel oil had a big share of the energy market. This resulted in an easy and cheap penetration in the solid fuel market as far as this fuel was used for heating purposes.

The above illustrates how the characteristics of the industry and the availability of gas determine the potential of gas in this industry.

## PREMIUM USES FOR GAS IN INDUSTRY

Part of the energy requirement of industry is covered by fuel such as coal, oil, or gas. The major quantity of these fuels is used to produce heat. In many cases the heat produced is not required at one specific place in the factory, but at several places; therefore, the heat has to be distributed throughout the plant. There are, in principle, two ways to realize this distribution: the fuel can be brought to the places where heat is required and burned at site, or the heat can be generated at a central place from the fuel, a heat transport medium such as hot water or steam being then used to distribute the heat.

The characteristics of coal and oil as fuels led to frequent adoption of the latter system.

The characteristics of natural gas are completely different from coal and oil. It is a clean fuel that can be transported up to the places where heat is required, it produces heat in a simple and controllable way, and its combustion products do not pollute the work spaces or the products that come in contact with them. In case natural gas is used as a fuel, the necessity to apply a heat transport medium for heat distribution is absent in many cases.

The use of natural gas in direct-fired gas systems offers industry the benefit of a maximum efficiency with a minimum of capital expenditure.

The capital cost of individual gas-fired systems is usually less than the cost of a boiler and the gas distribution system is cheaper than steam piping.

The efficiency of a gas-fired system is usually higher; there are no losses from steam pipes.

Each individual process can have its own heating system; the energy supply system does not suffer from low efficiency at part-load operation. In addition direct gas-fired systems can cope more easily with sudden changes of load than in an indirectly heated system with a boiler. Direct gas-fired systems will, for instance, offer benefits in the case of drying processes used in the food, textile, paper, ceramic and other industries.

From the above it will be clear that it will, in a lot of cases, be necessary to re-orientate the thoughts about the energy supply to a factory to enable natural gas to show to full advantage.

This is even more so if natural gas is regarded as a source of both heat and electricity for an industrial plant. Natural gas is an ideal fuel for prime movers such as internal combustion engines and gas turbines. Maintenance cost is lower on these engines in the cases when a clean fuel, such as gas, is used. Most industries require thermal energy and electricity. A prime mover produces mechanical energy that can be converted to electricity and in doing so it produces heat normally labelled as waste heat. If this heat is produced near a place where heat is required for a process, it can be brought to good use. This led to the introduction of the concept of electricity generation at the industry's plant site with systems incorporating prime movers that are now known as total energy systems.

Provided that the ratio between the electricity and the heat requirement of an industry fits in with the ratio in which these energies are produced by a total energy system, really good overall fuel efficiencies can be achieved. On the other hand, the total energy system requires investments and gives rise to maintenance.

A careful study of all factors involved is required to come to the evaluation of the benefits of the use of this typical gas application. In a lot of cases quite a good pay-out time on investment could be shown. In Table III a survey of electric power generation, including total energy systems in use in Dutch industry, is given.

In relation to the total amount of electricity produced in the Netherlands by power stations, the amount produced in total energy systems is small.

This leads to a consideration of the factors that influence the penetration of natural gas in the applications mentioned above. In the following section some additional premium uses of gas will be indicated. A complete survey of all premium uses, their importance, and applicability under different circumstances can hardly be given. The gas industry is aware of the importance of a proper utilization of the fuel it supplies and promoted this by informing its industrial customers about new developments.

## FACTORS THAT INFLUENCE THE USE OF GAS IN INDUSTRY

The cost of energy as part of the total cost of the product of an industry varies from industry to industry. The iron and steel industry, part of the chemical industry, and the ceramic industry have high energy costs ranging from 15-20 per cent. of the cost of their products. Others, such as the automobile industry, have energy costs that present only a few percent of the product. The higher the energy cost, the greater the need of efficient fuel utilization and the sooner investments will be made to optimize fuel utilization and vice-versa. This is demonstrated by the rather restricted use of total energy systems in industry. The capital available is spent on investments that decrease the production costs stemming from sources other than energy, in those cases where energy costs are low.

Another factor that influences the penetration of natural gas in certain applications in industry is industry's estimate of the reliability of special gas-fired installations.

At the beginning of the introduction of natural gas in the Netherlands, there were certainly doubts in industry concerning the safety of supply and use. Since 1965 the gas industry in the Netherlands has shown that, under all conditions normally met, gas supply to industry is reliable. Futhermore, the gas industry went to great pains to ensure that its customers use the gas

TABLE III

**Electric Power Generation, Situation in the Netherlands
on 1 January 1971**

| | MW | % of total | Number of generators | Output/ generator |
|---|---|---|---|---|
| Public utilities (of which back pressure generation in conjunction with town heating) ± | 8910.5 | 83.16 | 128 | 69.6 |
| Industrial steam plants: Condensing machines | 192 | 1.79 | 23 | 8.3 |
| Extraction condensing machines | 581 | 5.42 | 38 | 15.3 |
| Back-pressure machines | 572 | 5.34 | 124 | 4.6 |
| TE-Systems Gas turbines | 243.5 | 2.27 | 15 | 16.2 |
| Gas piston engines | 12.7 | 0.12 | 30 | 0.424 |
| Peaking power gas turbines | 189.5 | 1.77 | 6 | 31.6 |
| Gas piston engine/ generator sets | 13.3 | 0.13 | 39 | 0.34 |
| | 10714.5 | 100 | 403 | 26.9 |

safely. Regulations have been made concerning the construction of supply
systems for natural gas on the customer's premises, the design of equipment
such as burners to be installed, and the operation of this equipment.

As a result of the attitude of the industry towards safety and the actions
of the gas industry, including assistance to the industry in the engineering
stage of projects and inspection of safe operation of installations, the safety
record for gas-fired installations in the Netherlands is good.

The use of natural gas frequently leads to product improvement and
to an increase of production rate. In this case it is not the specific fuel consumption only that governs the decision to use gas but the evaluation of
additional benefits as well.

Product improvements are normally caused by the better temperature
control possible when gas is used as a fuel and a precise control of the atmosphere
around the products to be treated in case it comes in contact with the combustion gases.

The brick industry switched to the use of gas because the percentage of
reject material after firing dropped from 15 to 5 per cent. Furthermore, the
heat consumption was reduced 10 per cent and is now 550 M cal/ton of
bricks produced. Artificial drying of green bricks became feasible because
combustion gases of natural gas could be used directly for this purpose,

eliminating the necessity to invest in complicated indirect-fired heating systems.

There are other processes where production rates could be increased, such as the production of cast iron in cupolas where, apart from the replacement of highly priced coke by cheaper gas, a production rate increase of around 30 per cent. could be obtained.

A factor that influences the penetration of natural gas in industry is the method of heat transfer to be used to heat the product in directly-fired processes. It is a well known fact that normal natural gas flames enable a smaller portion of heat transfer to take place by radiation than is possible with oil flames.

Traditionally in some industrial processes heat transfer by radiation is important or thought to be essential. Special equipment can be installed to enable heat transfer by radiation in gas-fired installations. Before taking that decision, it is of importance to investigate whether heat transfer by radiation is all that important for the process or whether the process can be altered in such a way that heat transfer by convection, i.e. the more natural type of heat transfer, for gas flames can be used.

Glass melting tanks are frequently labelled as installations where radiant heat transfer is essential. Our experience with conversion of glass tanks indicates that normal non-luminous flames can give good performance, provided flame shape and furnace atmosphere are kept well under control.

Changes in the process and the installation for metal re-heating make gas a profitable fuel for processes that were traditionally based on radiant heat transfer. Installations using rapid heating techniques by high-speed convective heat transfer have a performance superior to the traditional equipment used.

Nowadays much attention is given to environmental problems present in certain areas. This can result in a growth of the share of natural gas in the energy market as long as sulphurous oxides are the limiting factor for the use of other fuels. Practically all combustion processes produce nitrogen oxides. Though the use of natural gas gives a lower nitrogen oxide production than the use of other fuels, it gives no final solution to the problems of the side-effects of energy production, such as pollution of the enviroment. On the other hand, it can and will give solutions to problems of industrial waste disposal.

The major problem with waste incinerators is the control of gaseous emissions. Gaseous emissions that are objectionable come from other processes as well. Natural gas is an absolutely unique fuel for the fume incineration process.

## THE USE OF NATURAL GAS IN BOILERS

The use of natural gas for boilers in industry is not an optimum application of this fuel and therefore its competitive position in face of other fuels is weak and determined mainly by its price. On the other hand, boilers specially designed to use gas as the only fuel are better hot water or steam generators than their solid or liquid fuel fired counterparts.

The burner installation of a gas-fired boiler is simpler. As the turn-down ratio can be made bigger with gas burners than with oil burners, fewer burners of bigger capacity, resulting in simpler and cheaper burner installations, can be used.

The room required to burn gas is smaller than to burn the same amount of oil, furnace volume can be smaller, again leading to lower installation cost.

The combustion gases of gas burners are free from soot, ash, and mostly free from sulphur. Convection parts of boilers stay clean; provisions for cleaning the boiler can be eliminated and the passages in the convection part can be narrower than in cases where solid fuel or fuel oil is used. This leads to compact boiler design.

The absence of sulphur results in lower dew points of the flue gases. The temperature of the flue gases at the exit of the boiler can be kept lower, resulting in a higher boiler efficiency.

Apart from the aspects that influence the cost of steam generation in boilers specially designed for gas, there are aspects that are of value for any gas-fired installation, such as the absence of fuel storage and fuel preparation.

If the boiler has to be designed as a dual-firing unit, many of the advantages mentioned above are lost.

## THE USE OF NATURAL GAS AS A FEEDSTOCK

In the Netherlands, natural gas finds or will find application as a feedstock for the chemical industry, producing ammonia, methanol, carbon black, and, on a smaller scale, hydrogen, inert gases, and similar products.

The type of chemicals that can be produced from natural gas depends on the chemical composition of the gas and the presence of higher hydro-carbons which increase the possibilities of natural gas as feedstock. The presence of impurities such as sulphur will make it a less interesting raw material.

Optimum results can be obtained by the industry if a constant gas quality can be guaranteed. The plant can be designed to work under optimum conditions for that gas quality. However, if a constant gas quality is essential for the chemical industry, it is an advantage to connect it to one specific gas field and not at the end of a complicated grid incorporating several injection points of gases of different origin. It leaves the gas company free to transport gases of the same or slightly varying combustion characteristics, but within relation to what is acceptable for the chemical industry, widely varying composition through its grid.

Another advantage of the location of a chemical industry near the gas field is that gas under higher pressures than normally supplied to the industry

can be delivered without unduly high cost for the industry and the gas company
   The chemical industry produces chemicals in bulk such as ammonia
and methanol which are profitable to the gas company because of the constant
high load factor and the relative low supply costs per cu. m resulting from mass
consumption from few sales points.
   Whether the chemical industry can be regarded as a premium application
for natural gas depends on factors that lie partly beyond the sphere of influence
of gas companies. These are technological developments in plant design, in
the transport of chemical products, and the economical and political
conditions in other areas where gas is available.

## CONCLUDING REMARKS

   The industrial market is an interesting one for natural gas because of its consta
fuel requirement over the year. Furthermore, it can in some instances be
supplied on an interruptible basis to compensate for the high gas consumption
in the domestic market during winter, thus producing a higher load factor on
the gas transport system.
   In order to ensure that optimum results of the gas utilization are
obtained for both the user and the gas supplier, technological changes in
energy generation and distribution have to be considered; the industry has to
adjust itself to a new fuel with special characteristics.
   This can be achieved by the industry in close co-operation with the gas
suppliers, who will, on the other hand, help and stimulate the appliance industry
to produce modern gas-using equipment.
   The cleanliness of gas and its combustion products, its high combustion
efficiency and its easy controllability will certainly prove to be of great value
in an industrial world that has to be careful with its wastes, including waste
energy.

## DISCUSSION

*J. A. Buckley* (Gas Council) said it was a privilege to open the discussion
on Mr Hijszeler's authoritative review of the rapid development of natural
gas into the Dutch energy market. It was particularly interesting also to
compare the developments in Holland as described with those in the U.K.
There were a good many similarities and many of the developments or
policies described could equally have been written as part of the
development of the U.K. gas market over the past four years.
   A similar-shaped chart to Fig 1 could have been shown for the U.K., but
with the starting point as 1967−68 and with projected penetrations of
total market around 15−16 per cent by 1975−76, as compared with
Mr Hijszeler's 40 per cent.
   Mr Hijszeler, however, warned them indirectly that global figures could
be misleading and that a sizeable chunk of the total market was not
substitutable by natural gas. So also in the U.K., excluding fuel used by

other fuel industries, transport, and agriculture, they planned to penetrate the domestic market by 40 per cent, the commercial market by 23 per cent, and the general industrial market by 19 per cent by 1976. This pattern was not quite apparent in Holland from Fig 2, unless under the heading "Public Distribution" there were included appreciable amounts to industry and commerce. Perhaps Mr Hijszeler could comment on this. There was, however, one apparently fundamental underlying difference in the two developments. One got the impression from reading the paper, with its references to price variations in quantity sold and in use, as under-boiler fuel and in other places, that pricing policies were not combined with planned and limited sales objectives and control. It was appreciated, and certainly shown clearly in other papers, that gas available and contracted for was, so far, much greater in Holland than the U.K. and this too was no doubt a factor when comparing, for example, the quantities of gas used for electricity generation in the Netherlands and the U.K.

Nevertheless, with a large but limited availability both of supply and economic transmission, it had seemed essential to them to plan their marketing in such a way as to tailor availability to supply and transmission (including storage) capacity so as to sustain the premium domestic and other loads over a long-term time-scale. For this reason, it was a fallacy to consider projected demand by any kind of extrapolation over previous periods and in the U.K. in four years they had already gone through a complete cycle of slow growth − rapid growth back to slow growth in contracted sales for the bulk industrial markets. It would be interesting to hear from Mr Hijszeler whether there was a firm plan for future natural gas sales that was projected and worked to or whether the price mechanism was the main effective regulator. Likewise with transmission capacity. Both countries employed the interruptible supply technique referred to. Could he ask Mr Hijszeler whether the conditions of interruptibility referred to provided the flexibility in operating the system at the constant load referred to?

Mr Hijszeler referred to the importance of premium markets in general, and the domestic market in particular, and the combination mix of markets for optimum results. They agreed. In four years they had reached a position where more than half the supply of natural gas was available for domestic use on a peak day's requirement, as against about one third of the total gas to domestic on an annual basis − a situation very different from manufactured gas operations. This difference between availability peak to availability annual was planned to increase further in favour of the premium domestic load.

On uses in industry, there were many similarities in the developments in both countries.

A similar profile of numbers of industrial firms and their use of gas was apparent here and they too found it desirable to undertake extensive marketing research into the establishment of fuel use, processes, premium loads, and competitive prices. Another common factor was the importance paid to safety and much of the work of their Midlands Research Station had been directed to improving the safety and performance of natural gas and dual-fired industrial plant.

*C. Hijszeler* replied that "Public Distribution" in Fig 2 did, as Mr Buckley presumed, include sales to smaller industrial and commercial users. Gasunie direct industrial sales were generally to users consuming over 2 million cu m/year, smaller users being supplied through the public distribution system.

Initial sales objectives required a rapid penetration of the industrial market and therefore for natural gas to be used as a boiler fuel in competition with heavy fuel oil. Present sales objectives required a more selective approach and they were concentrating now on the premium markets with corresponding adjustments in selling price.

On the question of flexibility and interruptibility, the gas distributor had two options. Either the distributor had a limited number of very large interruptible contracts with power stations, the interruptibility of which was easy to control, or the distributor had a large number of interruptible contracts which were less easy to control, but the arc of which would give a higher load factor than in the case of the few large interruptible contracts.

*J. M. Soesan* (Conch Methane Services Ltd) asked whether natural gas was likely to be used increasingly as an automotive fuel in urban areas or in transport fleets.

*C. Hijszeler* said that natural gas was an ideal automotive fuel and was already being used for this purpose in some places. In the Netherlands, however, the tax regulations were such that this use of natural gas was not economically viable and there was, therefore, no incentive to develop it.

*A. Cluer* asked Mr Hijszeler to comment on the use of nuclear power to produce hydrogen from water and then to use the hydrogen as fuel. Hydrogen combustion gave no pollution problems and its transportation was said to be cheaper than electricity. Hydrogen could be transmitted through existing natural gas networks.

*C. Hijszeler* thought this was an interesting development and the use of hydrogen as a means of transporting energy had much to commend it. He believed that as a form of energy hydrogen costs were higher than natural gas costs but lower than electricity.

*Miss M. P. Doyle* (Esso Petroleum Co. Ltd) said that press reports had recently indicated a revaluation of the availability of Groningen gas, resulting in a cutback of industrial sales. Was this the result of price distortion relative to other fuels, or too great a demand?

*C. Hijszeler* replied that the development of Groningen gas was rapid and its price was related to heavy fuel oil price, but with a ceiling. With the large increases in fuel oil prices in 1970–71, this ceiling was passed, resulting in a greatly increased demand for gas.

The reports of a revaluation of the Groningen field were incorrect.

*R. M. Crockett* (Esso Europe Inc.) said that the reappraisal of the annual production of the Groningen field down from 100 milliard cu m to

85 milliard cu m had nothing to do with the reserves available. It was related to the technical development of the field.

G. *Laading* (Royal Norwegian Council for Scientific and Industrial Research) asked Mr Hijszeler to comment on the nitrogen content of Groningen gas. Did Gasunie remove the nitrogen or was the design modified?

C. *Hijszeler* said that Gasunie did not remove the nitrogen from Groningen gas, although, of course, the presence of 14 per cent nitrogen meant modifications in burner design.

In the Paris area, at the interface of Lacq gas and Groningen gas, Gaz de France removed the nitrogen, but this was the only instance.

# Natural Gas and the Environment –
# Before Combustion

By D. H. NAPIER

*(Department of Chemical Engineering and Chemical Technology,
Imperial College, London SW7 2BY)*

## INTRODUCTION

In assessing the interaction of fuels with the environment, the selection and application of entirely objective criteria is often inhibited by traditional impressions, which may be true only in part. Thus the winning of coal is recognized as dangerous and dirty and also disfiguring to the countryside; with combustion of coal are associated various forms of pollution. While these criticisms are true, new technology has gone some way towards ameliorating these problems. In the case of electricity, a refined form of fuel, it is clean in use and has the advantage of flexibility in application. Under further examination it is less attractive in that it is costly, presents a hazard to persons and to property; pollution produced at generating stations must also be considered. In considering gas as a fuel assessment one may well commence from impressions of the odour of the old type of gas works or concentrate upon the danger of explosions.

Consideration will be given here to some of the properties of natural gas and to likely effects that these may have on the environment. Such effects may arise in two ways:

1. From the normal processes of winning, purification, storage, and distribution of the fuel.
2. From fault conditions in any of these processes.

A major consideration in all of these aspects is the safety of processes and situations, as well as the ability to predict the extent of hazards, thereby leading to the design of reasonable and effective precautions. In these respects experience has to be gained in handling and using a new material, but this experience must not be bought at the price of disaster.

In considering the properties of natural gas, it is appropriate to compare it

TABLE I

**Effect on the Environment of Some Natural Fuels**

| Process | Solid Fuel | Liquid fuel | Nuclear fuel | Natural gas |
|---|---|---|---|---|
| Winning | Pit heads Spoil tips Subsidence Effect of open-cast mining | Drilling rigs and well heads Restricted activities in their vicinity | Health hazards to those involved in mining operations | Drilling rigs and well heads Restricted activities in their vicinity |
| Treatment | Washing and screening plants producing spoil and waste liquor | Large refineries often linked with petrochemical complexes. Massive intrusion into the country-side | Separation plants requiring careful siting to minimize the danger of release of radioactive material Disposal of radioactive waste | Impurity separation and disposal of impurities |
| Storage | Piles on open ground that may self-heat Coal types require separate storage | Above-ground tanks presenting fire and explosion hazard | Quantities involved do not render this a major problem | Above-ground and in-ground storage presenting fire and, to a lesser extent, explosion hazard |
| Distribution | Unit basis - mainly rail and road | Rail, road, and pipeline | Special containers - high risk involved Fuel has to be returned for processing | Pumping through pipelines - usually concealed |
| Effect of leaks | Negligible | Fire and explosion hazard | Can be extensive and long term | Fire and explosion hazard |

with other gaseous fuels. However, in the light of the fundamental differences between gaseous and solid and liquid fuels, general comment only will be made rather than detailed comparison.

The brief general discussion of fuels is commenced by setting down in Table I some of the more obvious aspects of interaction with the environment prior to combustion. Generally, the winning of fuels is a hazardous occupation involving considerable numbers of men at risk. This is exemplified by coal mining, where, in a declining industry in 1968, there were 116 fatalities and nearly 870 seriously injured, while in 1970 the respective statistics were 91 and 650. The fuels compared in Table I are used substantially in the form in which they are found. The effect of the processes noted is in each case increased by the scale of operation both in respect of hazard and impact on the environment.

## PROPERTIES OF NATURAL GAS

Some properties of natural gas are compared here with those for other gaseous fuels and first, in Table II, some appropriate physical properties relating to liquefaction are given. In order to give definite values, the properties of methane, propane and butane are given rather than of natural gas and liquefied petroleum gas (LPG). Liquefaction characteristics of town gas are not appropriate.

### TABLE II

### Liquefaction Characteristics of Fuel Gases

| Gas | Sp.Gr. of gas 288.3K 1 atm Air=1 | Sp.Gr. of liquid | Boiling point K | Melting point K | Critical temper- ature K | Critical pressure atm | Gas/Liq. expan- sion ratio |
|---|---|---|---|---|---|---|---|
| Town gas (G.4) | 0.475 | | | | | | |
| Natural gas | 0.584 | | | | | | |
| $CH_4$ | 0.555 | 0.425 | 111.7 | 90.6 | 190.7 | 45.8 | 650 |
| $C_3H_8$ | 1.547 | 0.585 | 230.8 | 85.3 | 368.6 | 43 | 320 |
| $C_4H_{10}$ | 2.071 2.06 | 0.601 | 272.5 | 138 | n- 426 iso-407 | n - 36 iso-37 | 230 |

The values in Table II show clearly the relative difficulty of liquefying natural gas and the relative ease of so doing LPG, a matter of importance in consideration of storage and bulk transport. The specific gravity of these

TABLE III

Some Combustion Characteristics of Fuel Gases

| Property | Units | Nat. gas | Town gas (G.4) | $CH_4$ | $C_3H_8$ | $C_4H_{10}$ |
|---|---|---|---|---|---|---|
| Calorific value | Btu ft.$^{-3}$ | 1020 | 500 | 1012 | 2520 | 3106 |
| Wobbe number | | 1330 | 730 | 1370 | 2076 | 2219 |
| Flammability limits | % in air | 5 - 15 | 4 - 40 | 5 - 15 | 2.2 - 9.5 | 1.9 - 8.5 |
| Maximum burning velocity | cm s$^{-1}$ | 35 | 80 - 120 | | 46 | 40 |
| Air required for complete combustion | ft$^3$/1000 Btu | 9.4 | 8.8 | 9.55 | 9.62 | 9.65 |
| Diffusion coefficient | cm$^2$s$^{-1}$ | | 0.611 for hydrogen | 0.185 | | |
| Auto ignition temp. | $^\circ$C | 705 | 593 | 537 | 468 | 405 |
| Volume of flue gas | ft$^3$/1000 Btu | 10.4 | 10.2 | | 9.2 | 9.1 |
| Vol. of carbon dioxide | ft$^3$/1000 Btu | 1.00 | 1.01 | | 1.2 | 1.3 |
| Sulphur content | g/100MJ | 0.03- 0.06 | 0.08-6 | 0 | 0 0.40 (comml) | 0 0.41 (comml) |

gases is not only of interest in relation to storage and distribution but also in relation to leaks and the likelihood of layering. Natural gas, like town gas, will layer at roof or ceiling level, LP gases will collect in cavities.

In Table III some of the combustion properties of natural gas are set out against other fuel gases. Some of these values have an effect on the environment before combustion of the gas in respect of design considerations. For example, while natural gas requires a larger volume of combustion air than does town gas for equal volumes of fuel, comparison of the gases on the basis of heat release shows that the volumes are approximately equal. On a similar basis, the volumes of flue gas and of carbon dioxide are very similar. These values bear on the problem of modified air inlets and ventilation which have often been overstated in comments on the conversion programme from town gas to natural gas. In many cases the occasion has offered an opportunity to correct faults of long standing.

A major difference between town gas and the other gaseous fuels lies in the fact that it contains a considerable proportion of free hydrogen. Thus the ability of hydrogen to diffuse readily will be emphasized in considering small leaks from pipes containing town gas, but traditionally this has been a wet gas, so that the effectiveness of small leaks is minimized. The amount of gas diffusing through a porous pipe depends not only on the diffusion coefficient but also on the pressure drop across the diffuser. Transmission pressures for natural gas are higher at all stages, often very much so, than for town gas, and the natural gas is a dry gas. In both cases, however, diffusion coefficients are relatively high, so that dispersion in the atmosphere will be rapid; clearly, leaks of any gas into an enclosure are subject to different considerations and presents a major hazard. LP gases are not extensively piped and both diffusion and gravity result in slow dispersion of material from leaks.

The effect of hydrogen on the behaviour of town gas is further illustrated by the combustion properties. The flammable range is wider and the burning velocity higher for town gas as compared with natural gas or LPG. Further, the minimum energy of ignition for LPG and natural gas (0.3 mJ) is higher than for town gas (hydrogen 0.02 mJ). Again, a balanced judgment on safety must be made, in that accidental ignition for natural gas and LPG is more difficult but reliably designed ignition may also be so.

The pressure developed at a given time in an explosion in a given volume is proportional to the burning velocity [1]. It is the rate of pressure increase rather than the small variations in total pressure attainable, as indicated by adiabatic flame temperature, that determines the violence of explosions. Pressure relief usually arises from damage preceding the development of maximum theoretical pressure. From these considerations it would be expected that the effect of natural gas explosions would be less damaging to building than that of town gas explosions.

Another major difference of manufactured gases from natural gas and LPG lies in the presence in the former of a considerable proportion of carbon monoxide. This presents a major toxic hazard. As far as is known, natural gas is non-toxic and danger would only arise from exposure to massive releases producing asphyxiation. The situation with LP gases is somewhat similar, although attention has been drawn to the possibility of a correlation between this material and cardiac disorders [2].

Coal gas has its own strong and characteristic odour. This feature is essential in a gaseous fuel and stenching agents have therefore been added to natural gas, LPG, and manufactured gases. The effectiveness of this is not above criticism and casual observation suggests that there are more people who fail to notice stenched natural gas than those who are unaware of coal gas. This, and accommodation to odours, are factors to be considered in the discussion of the safety and economic aspects of installing fuel detectors in buildings where piped fuel supplies are installed.

The toxic effects of fuels on man are of prime importance, but the effect on trees and plants must also be considered. Leaks from pipelines, particularly from old ones used at higher pressures with natural gas, have occurred. Deleterious effects on trees have been reported from Holland and more recently in North Wales and in Yorkshire [3]. The treatment suggested for trees bears some similarity to hyperbaric oxygen therapy for humans, namely, that of increasing the oxygen supply to the roots.

One other interaction between man and natural gas arises from the fact that it is handled as a liquid (LNG). Contact "burns" may be produced by touching the gas or its immediate container. Hypothermia can result from experiencing the low temperature of the surroundings of the gas storage. Both of the hazards would only be realizable under fault conditions.

## FACTORS RELATING TO THE ORIGIN OF THE FUEL

The availability of natural gas at high pressure and at a relatively high degree of purity are factors strongly in its favour. Manufacturing plant is not necessary and the problems associated with mining are absent. In the U.K. much of the natural gas is obtained from offshore wells. Operational difficulties arise therefrom, but interaction with the environment is minimal. Further supplies are brought into the U.K. from the Sahara via Arzew. In both cases the effect on the environment arises from the siting of the terminals, the favoured choices for which are near the coast and remote from conurbations. Dependence on land-based supplies might alter this situation; exploration has not usually been welcomed by those close to it.

Once the gas has been brought to the surface, purification must be undertaken; the processing will vary according to the source of the gas. Gases from different sources contain differing quantities of sulphur compounds (hydrogen sulphide, mercaptans, and carbonyl sulphide), and processing is required both to dry and to sweeten the gas; the former is necessary, among other reasons, to avoid the formation of hydrocarbon hydrates [4]. Various processes [5] have been designed for purifying the gas.

Methods have been described for removing hydrogen sulphide and carbon dioxide using regenerative materials such as alkali carbonate and mono-, di-, and tri-ethanolamine. Water dew-point (to approx. 223 K) has been controlled by using high concentration glycol solutions or by adsorption drying. Water removal is necessarily undertaken at the well-head. On land-based installations the problem of water disposal arises and it has usually been solved by disposal at sea or in waterways. Recently, the suggestion [6] has been made that in order to avoid problems with frozen waterways, the water from the Groningen field will be injected back into the gas pay zone. Additional processing may include helium extraction (where this is present in workable concentrations) and nitrogen rejection. These may be achieved by low-temperature fractionation after cooling either by direct contact in a cascade arrangement or by auto-refrigeration, *e.g.* with a turbo-expander.

It is readily appreciated that purification plants are required alongside gas terminals. Waste chemicals (solid, liquid and gaseous) will be produced and a by-product or even twin-industry might develop, *e.g.* sulphur production

### TABLE IV
### Composition of Natural Gas and Some Manufactured Gases

| Gas | Percentages by volume | | | | | | | | | Calorific value Btu ft$^{-3}$ |
|---|---|---|---|---|---|---|---|---|---|---|
| | $H_2$ | CO | $CH_4$ | $C_2H_6$ | $C_3H_8$ | $C_4H_{10}$ | $CO_2$ | $N_2$ | $O_2$ | |
| Coal gas | 51.0 | 14.6 | 19.1 | 1.7 | 0 | 0 | 3.6 | 6.1 | 0.4 | 500 (sat.) |
| Reformed naphtha + $CH_4$ | 44.3 | 15.5 | 29.0 | 0.2 | 0 | 0 | 5.0 | 4.3 | 0.6 | 500 (sat.) |
| Total gasification of coal | 52.5 | 28.5 | 6.5 | | | | 8.0 | 4.0 | 0.2 | 335 (sat.) |
| Natural gas - West Sole | 0 | 0 | 94.1 | 3.2 | 0.6 | 0.2 | 0.5 | 1.2 | | 1041 (dry) |
| Natural gas - Hewett | 0 | 0 | 81.8 | 6.0 | 2.5 | 0.4 | 0.1 | 9.0 | | 1060 (dry) |
| LNG - Algeria | 0 | 0 | 87.7 | 8.6 | 2.3 | 0.9 | 0 | 0.4 | | 1132 (dry) |
| Natural gas - Lacq | 0 | 0 | 82.1 | 3.3 | 1.0 | 0.7 | 11.6 | 0.2 | | 980 (dry) |
| Natural gas - New York supply | 0 | 0 | 94.5 | 3.3 | 0.7 | 0.3 | 0.7 | 0.3 | | 1049 (dry) |
| Natural gas - Moscow supply | 0 | 0 | 93.2 | 0.7 | 0.6 | 0.6 | 0 | 4.4 | | 1006 (dry) |

at Lacq. Over a period of many years, statutory legislation has evolved in the U.K. that is effective in controlling the disposal of liquid and gaseous effluents that may be produced from factories and power stations (*e.g.* Alkali Works Regulations, Clean Air Act, Public Health Act, Clean Rivers Act, etc.). The requirements imposed for protecting the environment are becoming more stringent with the passage of time and they will soon be reinforced by the 1972 Deposit of Poisonous Waste Act. Natural gas is potentially well-placed to meet all of these requirements, as also would manufactured gas; coal gas would have been in an increasingly difficult, and so less economically favourable, position. The generation of electricity from fossil fuel causes major problems (oxides of sulphur and (from coal) ash), as it does from nuclear stations (disposal of radioactive waste and old reactors).

After processing, gases of composition [7] shown in Table IV are transmitted through pipeline systems to the consumer. For purposes of comparison, the composition of three manufactured gases is given together with the calorific value each gas.

## STORAGE

Although high-pressure pipelines store large amounts of gas (line-packing), when fuel supply is discussed on a national scale considerable storage volumes are required; peak-shaving demands such reserves. Several types of storage are in use and these will be considered here.

1. Gas may be stored at high pressure in pipes ("bulletts"), in tanks or in oversize mains. Specifications for pipelines will be mentioned later; adequate experience exists [8] on the design and construction of pressure vessels so that no new problems can be envisaged. It may be that further consideration has to be given to the design of tank foundations and bases and to the effect of wind loading.

   Oversize mains can be fairly readily hidden from view and offer little other change to the environment than that occasioned by trench-digging. Storage spheres or other above-ground gas-holders are not readily hidden and rank in offence with petroleum storage.

2. Another way of providing storage is to keep the gas as LNG either above-ground or in-ground. The LNG storage facility at Canvey Island is well known and further liquefaction and storage will be available at Glen-mavis (Scotland) with yet others planned for South Wales and Partington. Methods and materials for above-ground storage are both topics under active consideration. The necessity for care in design and siting is emphasized by the Cleveland disaster of 1944, when 128 people died and some 400 were injured; property damage was estimated at nearly $7 x 10^6 [9]. Attention has been drawn to many features of the disaster, among them the close proximity of the site to a large conurbation, the lack of dykes and adequate bunding, the presence of sources of ignition,

the choice of tank material. Materials that do not embrittle too severely are essential for such tanks; such materials include nickel, aluminium, copper, and alloys containing high percentages of nickel. The values [10] given in Table V illustrate the manner in which impact strength falls with temperature.

### TABLE V

**Impact Performance of Cooled Nickel Steels**
**Tested by Charpy K Impact (Keyhole Notch) Test**

| % nickel | 173K | 123K | 73K |
|----------|------|------|-----|
|          | ft - lb. | ft - lb. | ft - lb. |
| 0        | 2    | 0    | 0   |
| 2        | 15   | 0    | 0   |
| 3.5      | 38   | 25   | 0   |
| 5        | 45   | 31   | 8   |
| 8.5      | 35   | 30   | 20  |
| 13       | 32   | 31   | 30  |

Humbert-Basset [11] has reported on the successful use of steel containing 9 per cent nickel, 0.1 per cent carbon, and 0.04 per cent aluminium; this material gives the necessary ductility, but does not require stress relieving, as does 9 per cent nickel steel. Austenitic stainless steel and copper are both suitable but are too expensive.
Gibson and Walters [11] have described the use of double skinned tanks having an inner skin of the face-centred cubic aluminium or stainless steel or the body-centred cubic 9 per cent nickel steel. The outer skin is then fabricated from a cheaper material with adequate insulation between. Tanks have been built to contain 20,000 tonne and designs are being evolved for 40,000-50,000 tonne storage.
The development of pre-stressed concrete tanks continues. Humbert-Bassett [11] reported the use of tanks 15 m in diameter, 13 m in height (approx. 2000 m$^3$) with 0.40 m Perlite and a gas-tight Invar skin. A tank of about 90 m diameter (90,000 m$^3$) has been built on Staten Island and Closner [11] forecasts that tanks up to $4 \times 10^5$ m$^3$ could be produced.

3. Design of storage tanks must take account of economics and the effect on the environment, as well as of safety and reliability. The cost of storage for LNG is likely to be high and the safety demands to be stringent. Methods of storage have to be kept under constant review. In-ground storage seems to be a very attractive proposition, wherein the roof is the only fabricated structure. The storage is generally unobstrusive and site areas are about one half of that for equivalent aboveground tanks. However, the use of such storage appears to be confined to

base load operations where high boil-off rates can be tolerated.
Comparative boil-off rates for above-and in-ground storage have been
quoted as 0.04 per cent and 0.1 to 0.3 per cent per day respectively [11].
This feature has led to the abandonment of some in-ground storage,
*e.g.* Hackensack and Hopkinton in the U.S.A. Additional concern arises
from the thermal stresses in both rock and frozen soil and the cracks
that result therefrom.

A somewhat similar method of storage that is used commercially for
LPG and experimentally for LNG is the utilization of mined caverns.
The cost of excavation is low, but those of insulation and lining are
high. While in-ground storage seems to offer less hazard and less offence
to the environment, there remains a need for development of the
technology involved.

## DISTRIBUTION

High-pressure bulk transmission of natural gas at between 500 and 1000
psi is undertaken in pipelines for which specifications have been drawn up
by the Institution of Gas Engineers [12], dealing with materials, pre-treatments,
wall thickness, laying and welding and test procedures. Some of the features
appropriate to this discussion are given in Table VI.

## TABLE VI

### Some Aspects of High-Pressure Pipeline Specification

| | |
|---|---|
| Steel | BS 3601-5   API 5L, 5LX, 5LS. |
| Finish | De-scaled and phosphate·treated |
| | Inside pipe        : red lead |
| | Outside pipe      : bitumen or coal tar enamel reinforced with |
| | glass fibre PVC tape or plastic sheath |
| Wall thickness | $K = \dfrac{PD}{2fs}$   in. |
| | P -    maximum working pressure lb f in $^{-2}$ |
| | D -    outside diameter of pipe     in. |
| | s -    specified minimum yield strength lb f in $^{-2}$ |
| | f -    factor, 0.72, based on minimum wall thickness, maximum |
| | working pressure, and grade of steel |
| Testing | Air at 100 lb f in $^{-2}$ for at least 24 hr. |
| | Hydraulic tests at 1½ times maximum working pressure |
| Protection | Taping or hot flood coating at joints |
| | Inspection for holidays − current drainage survey |
| | Cathodic protection − usually 0.01-0.02 mA ft $^{-2}$ |

In the context of the environment the section of the recommendations that
deals with maximum permissible pressures in different areas is noteworthy. A
summary is given in Table VII.

### TABLE VII
### Maximum Permissible Working Pressures

| Area type | *Population density and description | Max working pressure lb f in$^{-2}$ | f |
|-----------|-------------------------------------|-------------------------------------|------|
| R | 1 person/acre rural area | 1000 | 0.72 |
| S | 1 person/acre residential property, schools, shops | 350 | 0.55 |
| T | high density of population, heavy traffic, numerous underground services | 100 | 0.40 |

*Population density is the average for each mile of pipeline in a strip centred on the pipeline and of width ten times the minimum separation distance of pipelines from buildings.

The values given in Table VII lead to the requirement for recommended separation distances given in Table VIII.

### TABLE VIII
### Minimum Distances of Pipelines from Buildings

| Limits of Outside Diameter of Pipe in. | Maximum Working Pressure (lb f in$^{-2}$) of | | |
|---|---|---|---|
| | 100 | 350 | 1000 |
| | Requiring separations (ft) of | | |
| 6 5/8 | 10 | 15 | 75 |
| 6 5/8 - 12 3/4 | 10 | 40 | 100 |
| 12 3/4 - 18 | 10 | 60 | 125 |
| 18 - 24 | 15 | 80 | 160 |
| 24 - 30 | 20 | 100 | 200 |
| 30 - 36 | 20 | 125 | 250 |

Exemptions from proximity and pressure limitations are allowed in special circumstances in Type S areas on the grounds of extra protective measures being taken such as:

1. The provision of steel sleeving
2. Periodic hydraulic re-testing
3. Interposing a protective barrier between the pipeline and adjacent property.

Special protection is recommended for road and rail crossings and at any place where interference with the pipeline may be anticipated.

The impact of these recommendations is twofold. First, they delineate guidelines for the siting of pipelines; secondly, they will be effective in shaping the pattern of future development. They therefore have a direct influence on the environment.

There is a body of statutory legislation bearing on the route of pipelines. Area Gas Boards are thus enabled to install distribution networks. They may proceed in two ways under the 1948 Gas Act in laying mains through private land, namely, either by negotiating easements for laying, repair, and maintenance or by compulsory purchase. The 1950 Public Utilities Street Works Act, the 1949 Special Roads Act, and the 1959 Highways Act define the powers available to Area Boards for pipelaying.

Safeguarding of the environment has been provided for in many ways, *e.g.* controls relating to operating in various areas:

1. Agricultural land (1950 Diseases of Animals Act)
2. Rivers, canals, and waterways (1948 River Board Act)
3. Railways (1950 Public Utilities Street Works Act)
4. Forests and woods (1951 Forestry Act)

Agreement with landowners allows for a permanent easement of 20 ft in which no building or construction work is permitted. In addition, 10-ft support strips on either side, in which work may only be carried out with the permission of the Gas Council, give an added degree of safeguard.

The extensive precautions taken to protect the environment in both the aspects of safety and amenity are supported by adequate engineering design requirements. Thus, isolating valves must be installed at distances of less than 10 miles and the spacing in built-up areas is decreased.

The situation arising from the extensive network of old pipelines used for wet gas at lower pressures, much of which has open-socketed joints with hemp and lead seals, is unsatisfactory. Drying of the seals by the dry natural gas has caused leaks and necessitated checking and treatment of joints. The reliability of the distribution system, which amounts to some 120,000 miles of main and several million joints [7] will improve as surveys and repairs are undertaken. Extensive renewal of medium- and low-pressure pipes will be necessary; in many cases, corrosion has already weakened pipes that are now being subjected to higher pressures. Another feature of old mains is that they usually contain considerable dust and debris. In order to minimize abrasion limitation of the gas velocity is necessary. For the worst cases a limit of 40 ft $s^{-1}$ is set; if the amount of dust is small, velocities up to 70 ft $s^{-1}$ may be tolerated[12]. It is unfortunate that natural gas has had to be supplied to the public through a system that had in some cases already exceeded and in others was approaching the end of its useful life.

A further aspect of interaction of the gas distribution system with the environment is the matter of noise pollution. Difficulties are likely to arise at pressure-reducing stations. An example of how these can be overcome is afforded by the action taken at the Ebbw Vale steelworks [13]. This was twofold:

1. By adequate use of sound insulating materials
2. By siting the station approximately 100 m from the nearest dwellings.

It has been suggested [11] that LNG could be distributed. Ivantsov, Livshits and Rozdestvensky have described a 56-inch diameter line operating at 75 atm, but they conclude that cheaper steels safe in operation at low temperatures are necessary for this method to be economically viable. Hoover has suggested that an insulated pipeline of 9 per cent nickel steel, austenitic stainless steel, or aluminium would be useful; polyurethane foam or polyvinyl chloride is suggested as the insulant. The transmission distance would be limited by the rise in temperature from, say, 110 K to 175 K.

## DISPERSION OF GAS IN THE ATMOSPHERE

Mention has already been made of the effect of small, long-term leaks on vegetation, but consideration must also be given to possible effects of the sudden failure of pipelines and to the lifting of safety valves. In both cases emission of natural gas should be of limited duration; in the first case flow will cease by the operation of stop valves and in the second case the relief will close when the excess pressure has been vented.

If the leaking gas (natural gas or town gas) is able to enter a confined volume, roof layer formation is likely to occur. This is a common hazard in coal mines [14] and can give rise to destructive explosions; it also occurs in domestic premises. With LPG there is a counterpart to this at floor level.

Expressed in more general terms, the failure of a high-pressure natural gas main may produce conditions for asphyxiation in the vicinity of the leak and a volume within which the gas composition will be within the flammable range. In both cases, limits may be estimated in conditions of high wind and for a quiescent atmosphere. In the case of natural gas, the concentration limit in the first zone will be from 100 per cent to about 50 per cent and the range in the second is from 5 to 15 per cent.

Estimation of the extent of dispersion has been carried out with acceptable degrees of success on the basis of work by Bosanquet and Pearson [15] and Sutton [16]. Concentration profiles may be computed under a variety of meteorological conditions, provided that a realistic estimate can be made of the effective height of the emission above the ground, *i.e.* the sum of the actual height and those additions arising from momentum and buoyancy effects.

Another type of failure system arises either from the handling of LNG as in the case of the failure of a transfer line or from the failure of storage. This latter may be from a major tanker collision at sea or the rupture of an above-ground tank. LNG will then be spilt on a surface warmer than itself and flash evaporation will occur for a period of up to a minute, followed by a subsequently lower rate of convection-controlled boiling. At low wind speed this latter remains constant at a regression rate of about 0.5 mm min.$^{-1}$. Gas so produced will either mix with the air or stratify [17], depending on the wind speed.

From experiments undertaken with spillages of LNG, empirical relations[18] have been obtained showing that the diameter of the pool is a linear function of time, and that this spreading persists until the liquid film breaks and bare patches of ground appear. On this basis the maximum diameter of the pool $\underline{d}$ max is given by

$$\underline{d} \ \text{max (ft)} = 6.25 \ W^{1/3}$$

where W is the weight of LNG in lb.

It is likely that the gas will inflame and pool burning will occur. The regression rate is given by Burgess and Zabetakis [19], as follows:

$$\underline{V} = V_\infty \ [1 - \exp(-hd)]$$

where d is diameter of the liquid pool (ft)

      h is a constant approx 1, if d is in ft.

      $V_\infty = 0.0030 \times \dfrac{\text{net heat of combustion}}{\text{sensible heat of vaporization}}$

Hottel [20] has given the rate according to

$$V = \frac{\sigma T_F^4 F}{\rho \Delta H_V} \ [1 - \exp(-kd)]$$

where $T_F$   is   flame temperature

    $\sigma$    is   Stefan-Boltzmann constant

    F    is   shape factor

    $H_V$   is   volumetric heat of vaporization

    k    is   opacity coefficient

    d    is   diameter

Some of these considerations are concerned with combustion but occur prior to designed combustion. They have clear effects on the environment; asphyxiation, fire, and explosion may result therefrom. A further effect arises from the fraction of heat produced during combustion that is released by radiation. Some comparative values [21] are given in Table IX, from which it will be noted that the radiation from a burning pool of LNG is low.

### TABLE IX

**Radiative Characteristics of Fuels**

| Fuel | Release as radiation % |
|------|:----------------------:|
| LNG | 23 |
| n-Butane | 30 |
| n-Hexane | 40 |
| Benzene | 38 |
| Methanol | 20 |
| Gasoline | 30 |

The sequence following a spillage of LNG consists of a short period (about one minute) of flash-evaporation, followed by gas production from a quasi-steady state. If ignition occurs, there will be a large momentary flash before pool burning sets in and flame stabilizes above the spillage [22]. Burning rates are then somewhat similar to those for gasoline. The fire may be extinguished by limiting the oxygen with inert gas addition or by careful use of dry powder. Too large an addition of dry powder increases the rate of gasification. Humbert-Basset [11] reported the extinguishment of a burning pool of LNG of area 50 $m^2$ in 12 s with 800 kg of sodium bicarbonate. In that the powder does not float and has limited throw (approx.50 m) a mixture of powder and foam (expansion 200) has been tried and found to be successful.

Spillages of LNG on water have also been examined [23], when an initial evaporation rate of 0.037 lb $ft^{-2}$ $s^{-1}$ was recorded which moderated after 20 s. Ice floes formed with confined spillages but not with unconfined. Average downwind concentrations may be calculated, but measured values of peak concentration were up to 20 times average values.

These spillages are somewhat similar to those for liquid fuels. Two main differences arise: first, greater control is required in fire-fighting; secondly, if ignition does not occur immediately, the chances of an explosion occurring are much greater with LNG than with liquid fuels.

## STATIC ELECTRIFICATION

The possibility of static electrification occurs in many systems where there is movement between phases[24] and charge separation can occur. Thus, dust moving in a pipeline will become charged, but it is unlikely that the dust will be present in sufficient amounts for energy to accumulate, so that at some stage in the system an incendive spark could be formed. Further, for dusty pipelines a limit is set on the speed of transmission for reasons of abrasion and this of itself will minimize the static charging.

If LNG is pumped through a pipeline, charging may occur, charges may separate at the pipewall (streaming potential) or if particles are suspended in the liquid from this cause (sedimentation potential). The amount of charge depends on the ion population in the LNG; the expectation would be for it to be low. Values quoted by Humbert-Bassett[11] support this: e.g. at 250 V, resistivity 5 x $10^{14}$ Ω $cm^{-1}$, at 500 V, resistivity $10^{15}$ Ω $cm^{-1}$.

It was found that after prolonged pumping this value rose and that sulphur-containing impurities reduced the resistivity. If comparison is made on this basis with petroleum products, it is above the range for white products, $10^8$ to $10^{11}$ Ω $cm^{-1}$, and well above that for the "safe" crude petroleum, $10^5$ to $10^7$ Ω $cm^{-1}$. Charging currents measured from pumping LNG through non-oxidizing pipelines are lower than for JP 4.

The hazard of the charged fluid will not be realized unless an incendive spark is produced in a mixture of natural gas with air within the flammable range. With a highly volatile fluid like LNG the likelihood of spaces near liquid surfaces being fuel-rich, is very high. The hazard of static charging in the LNG seems to be low, but the cautionary note [25] sounded in relation to the transfer of LNG by temporary connection from road tankers is undoubtedly justified.

## DISCUSSION

The effect of natural gas on the environment may be considered at three levels, namely:

1. Personal and domestic
2. Local
3. Large-scale and national

The impact of natural gas at the first level goes somewhat deeper than the relative trivialities of conversion. These are normally little more than inconvenience and dislocation of a short-term nature. It is, however, important in that a large percentage of the population of the U.K. have become directly involved. The occasion may call for modifications to air supply and ventilation arrangements with the bonus that servicing of appliances, often long overdue, is carried out.

The carbon monoxide hazard prior to combustion is removed by conversion from town gas to natural gas. The other hazards at level 1 relate to fire and explosion. The hazard of the fuel is usually inseparable from the effects of inadequate apparatus and the occurrence of fault conditions. By way of illustration, attention to use of liquid fuels in the domestic environment can be directed along two lines. First, the hazards of a well-engineered central heating system are low, although the necessity to pump the fuel to the burner and the requirement for on-site storage must be set against other advantages of liquid fuels. Secondly, attention may be concentrated on the free-standing heater, where often apparatus is poor, the context of use is incorrect, and whence a steady stream of fatalities results.

## TABLE X
### Sources of Ignition of Fires

| Source of ignition | Fires in dwellings in U.K. (%age) | |
|---|---|---|
| Ashes, soot | 2.0 | |
| Electrical sources | 34.2 | |
| (Cookers) | | (12.8) |
| (Heaters) | | ( 4.3) |
| Gas appliances | 8.2 | |
| (Cookers) | | ( 6.7) |
| (Heaters) | | ( 0.6) |
| Smoking materials | 8.4 | |
| Oil appliances | 9.0 | |
| (Heaters) | | ( 6.8) |
| Solid fuel appliances | 21.1 | |
| Children with fire | 4.6 | |

The advantages of electricity on the domestic scene are manifold, although cost may be high. It does, however, possess the major and obvious disadvantage that it is the only fuel that can cause electrocution; in addition, it also causes many fires. Some statistics for 1965 shown in Table X[26] illustrate the point to the extent that electricity emerges as the largest single source of ignition. While a full analysis of these figures and of those given in Table XI will not be undertaken here, the trend appears of a higher number of fires per $10^6$ appliances in the case of electricity as compared with gas. There is no reason to anticipate that conversion to natural gas will do other than improve the situation in relation to gas appliances.

TABLE XI

**Numbers of Appliances in 1966 in U.K.**

| Appliance | Electric | Gas |
|---|---|---|
| Cookers | $6 \times 10^6$ | $11 \times 10^6$ |
| Water heaters | $7.8 \times 10^6$ | $3.3 \times 10^6$ |
| Space heaters | $26 \times 10^6$ | $7 \times 10^6$ |
| Central heating | $0.6 \times 10^6$ | $0.8 \times 10^6$ |
| Fires/$10^6$ appliances | 280 | 129 |

At the personal and domestic level, poisoning by carbon monoxide before combustion can, by use of natural gas, be eliminated from consideration. There is then the matter of electrocution to set against that of explosion; in 1966 the Registrar-General reported 50 deaths by electrocution in the home (or 2.8 death per $100 \times 10^6$ therms of electrical energy supplied). In the same year it may be estimated that there was one death arising from gas explosions per $1000 \times 10^6$ therms of energy supplied. Clearly, the number of explosions in domestic premises involving town gas will be very much greater on the grounds of consumption than those involving LPG and liquid fuels. A case could be made out for more safeguards and a greater degree of control in the fitting and servicing of appliances.

The effect of natural gas on the environment at the second level relates in the main to terminals and storage. Power stations and gas terminals may be deemed equally offensive to view for those who visit their neighbourhood or are resident near them. Equally, they have little impact on those who never see them. This level of consideration merges into the large-scale and national. Power and energy have to be distributed, and the larger the power consumption the greater is the investment in the transmission system. The economics of underground cables are unfavourable as compared with pylons and overhead transmission lines. The disfigurement by them and by transformer stations is considerable and the hazard from overhead lines is not negligible. The majority of natural gas pipelines are buried, as are the control point; offence to the environment is thus minimized. Further, the amount of above-ground storage as compared with the period when town gas was in use has decreased. In view of the fact that natural gas is consumed at a multitude of locations and is sulphur-free, the effect on the atmosphere, as compared with the emissions from tall chimneys and cooling towers of the power station, is small.

One way in which natural gas may impinge on the local environment is in the matter of LNG tankers in port. The LNG will continue to evaporate so that it must either be re-liquefied, burnt, or dispersed. Fire and explosion could then only be avoided by imposing restrictions on access and movement more so than for petroleum tankers and stores. The whole issue of designated areas and the installation of flameproof switchgear comes up for consideration. This must be extended to aspects of dispersion already outlined in relation to spillages involving "risk-point" determination. It may well be that the re-thinking suggested [27] for the petrochemical industry is applicable in this case also.

Solid fuels have left permanent scars behind them in the form of spoil tips and, in some cases, the further effect upon water percolating through them. Distribution of solid fuel calls not for a different system, only a larger one. The protection of the atmosphere afforded by the Clean Air Acts has made the domestic solid fuel supply more complex in all respects. Very much the same considerations apply to liquid fuels.

A case has been made out on several occasions for the banning of gas installations in high-risk blocks. The Ronan Point disaster [28] illustrated not only what can happen, but that often there are contributory causes to such an event. Further, to imply that a gas explosion in a structure of this type was the only means by which such a disaster could happen, is to ignore bottled gas or volatile liquid fuels as sources of energy or even a pressure burst from a fault condition in a hot-water tank heated by an immersion heater.

On balance, it appears that natural gas offers no insuperable problems prior to combustion and those problems that it does pose are not new. Equally clearly, it helps in endeavours to protect and preserve the environment as compared with the effect of other fuels. It is to be hoped that every effort will be made fully to quantify every aspect of its impact on the environment so that its advantages may be exploited to the full.

## REFERENCES

1. Maisey, H. R.,*Chem. Process Engng,* 1965, 46, 526-35, 662-72.
2. Daily Telegraph, 22.12.70
3. Daily Telegraph, 25.2.72.
4. Backhurst, J. R., and Harker, J. H., *J. Inst. Fuel,* 1970, 43, 405.
5. Walker, J. E., and Webber, C.E., *Chem. Engng Progr.,* 1961, 57, Symposium Ser. No. 34, 50.
   Anon.*Hydrocarb. Process,* April 1971
6. Anon. *Oil Gas J.,* June 1971.

7.  "Report of the Inquiry into the Safety of Natural Gas as a Fuel".
    F. Morton. London, H.M.S.O., 1970.
8.  *e.g.* B.S. 1500, 1958-1965.
    B.S. 1515, 1965.
    B.S. 1501-1506, 1958. London, British Standard Institute.
    "Boiler and Pressure Vessel Code", 1968, ASME.
9.  Elliott, M. A., Seibel, C. W., Brown, F. W. Artz, R. T., and Berger,
    L. B. *Bur. Min. Rep. of Invest.,* 1946, 3867.
10. "Cryogenic Fundamentals", Ed. G. G. Haselden, Academic Press,
    London, 1971.
11. LNG 2nd International Conference, Oct. 1970. Paris.
12. Communication 674 A, 1970.
    Publication 674, 1967.
    Inst. Gas Engrs, London.
13. Rodd, P., *J. Inst. Fuel,* 1972, 44, 51-57
14. Bakke, P., SMRE Res. Report 164 (1959).
    Bakke P., and Leach, S. J. SMRE Res. Report 195 (1960).
    Perlee, H. E., Liebman, I., and Zabetakis, M. G., *Bur. Min. Rep.
    Invest.,* 1964, 6348.
15. Bosanquet, C. H., and Pearson, J.L., *Trans. Faraday Soc.,* 1936, 32,
    1249.
16. Sutton, O. G., *J. Roy. Meteor. Soc.,* 1947, 73, 426.
17. Bakke, P., Internat. Conf. of Directors of Safety in Mines Res.,
    1959, Paper 44.
18. Merte, H., and Clark, J. A., *Trans. ASME Ser. C.,* 1964, 86, 351.
19. Burgess, D. S., Strasser, A., and Grumer, J., *Fire Res. Abs. and Revs.,*
    1961, 3, 117.
20. Hottel, H.C. *ibid.,* 1958, 1, 41.
21. Tutton, R. C., "Safety in Handling Low Temperature Industrial
    Fluids", Symp. of SE Branch., Inst. Chem. Engrs, London, 1965.
22. Burgess, D. S., and Zabetakis, M. G. *Bur. Min. Rep. Invest.,* 1962,
    6099.
23. Burgess, D. S., Murphy, J. N., and Zabetakis, M. G. *ibid.,* 1971,
    7448
24. Napier, D.H., Inst. Chem. Engrs. Symp. Ser. No. 34, Inst. Chem. Engrs,
    London, 1971, p.170.
25. "Cryogenics Safety Manual", London, British Cryogenics Council,
    1970, p.72
26  Fry, J.F., FR Note No. 674, 1968, Fire Research Station, JFRO.
27. Arnaud, F.C., Conference Publication No. 74, London, IEE, 1971,
    p.145.
    Nixon, J., *ibid.,* p.150.
28. "Report of Inquiry into the Collapse of Flats at Ronan Point,
    Canning Town". London, HMSO, 1968.

## DISCUSSION

The discussion was opened by *C. Hijszeler* (N.V. Nederlandse
Gasunie), who commended Dr Napier upon his paper. It was particularly
important to incorporate LNG transportation into the paper, since LNG
transport would increase rapidly in volume over the next ten years and
it was important that no accidents should occur.

Although the industry did not consider natural gas unsafe, the public
did. If one looked at the accident records, one saw that most accidents did
not result from faulty appliances but from leakages in gas distribution
lines. Gasunie in the Netherlands had the duty of checking distribution
lines in industrial plants both with respect to design and construction.
Improvements in the quality of design and construction occurred as the
result of investigations carried out by Gasunie. What policies were being
adopted in the U.K. in this respect?

The transportation of electricity was creating great environmental
problems. Was Dr Napier familiar with work on the future development
of electricity transmission?

Dr Napier was asked if he knew of a method for the disposal of gas
odorants when accidentally spilled.

*Dr Napier* asserted that gas need not be unsafe. The Ronan Point
disaster was an example. Inside the building there was a rogue nut which
broke apart, allowing gas to leak into a kitchen in one of the flats. At the
public enquiry, attention was directed on the question of whether it was
possible to introduce a form of statutory control for appliance fittings.
It was seen to be unrealistic to attempt to exercise control over every
single nut. Gas undertakings in this respect were burdened with a very
serious problem.

With regard to transport of electricity, Dr Napier was horrified by
some of what he had heard, and as an example mentioned the super-
cooling of cables with liquid hydrogen.

The electricity industry moved in "phases". After fuel cells and
gasification, there came MHD but the next development was unclear.

Dr Napier could offer no solution to the problem of dispersal of
accidentally-spilled odorants, but proprietary materials were available that
ameliorated the situation.

*J. Wharton* (CEGB) reminded the audience that a sensible approach to
environmental problems was important. Criticisms by environmentalists
usually relied on exaggeration and were overplayed. *Dr Napier* agreed that
environmental problems could be distorted by over-emphasis.

*R. S. Hackett* (Gas Council) found nothing to disagree with in the paper,
since natural gas was shown to have so few ill-effects on the environment,
and commended Dr Napier upon producing an objective paper on a
difficult subject. While the U.K. gas industry as a nationalized industry
was over-sensitive to criticism, the industry did feel that it was doing what
it could, and the facts spoke for themselves.

Safety problems had always been with the gas industry. With greater volumes of sales, naturally the scale of leakage had increased. But this was not, as suggested in the paper, due to the change to natural gas. Leakage increased with the changeover to dry gas, which occurred when naphtha reforming superseded coal carbonization. Some areas of the U.K. had been using gas that was both water- and hydrocarbon-dry for many years.

The codes developed for town gas transmission in high-pressure pipelines were equally applicable to natural gas.

Most leaks that occurred were in low-pressure pipelines. Here, with the change from 8–10-inch water gauge with manufactured gas to 13–15-inch for natural gas, although the incidence of breakage was no greater, each time a break occurred the leakage might be greater.

The transmission mains were designed to operate at 67 Bar, so that the changeover to natural gas only allowed the Gas Council to operate these mains at design pressures.

As far as disasters such as Ronan Point were concerned, it was impossible to cover every eventuality, and indeed there were always people who would act in a misguided way.

*Dr Napier* apologized if the emphasis on natural gas creating greater leakage problems had displeased those attending from the Gas Council. As far as high-pressure mains were concerned, a rupture must occur sometime, and there was a need for knowledge of what happens upon rupture at high pressures.

Dr Napier reminded the audience that as a participant in the Ronan Point enquiry he had pointed out that the damage would have been as great if an LPG cylinder had been misused and leaked or if a man had washed his overall with petrol and such fuel had ignited and caused an explosion. The problem was not so much the gas as the method of construction of the building.

*R. Evans* (Gas Council) brought to Dr Napier's attention the comment on the diffusion of hydrogen and pointed out that with natural gas containing no hydrogen the principal component of diffusion was absent.

Whilst in some areas low-pressure gas distribution systems had been slightly increased in pressure, in many areas medium-pressure and high-pressure gas transmission systems were now carrying gas at lower pressures, taking advantage of the increased calorific value of natural gas *v.* town gas.

The reason that velocities were restricted in low-pressure pipes was not to combat erosion and abrasion from rust, as suggested, but rather to give the customer the best possible service and here sudden movements of rust would be unacceptable, leading to blockages of mains, pipes, meters, and appliances.

There was a great deal of difference between town gas and natural gas in an enclosed space and this was apparent from consideration of the differing gas compositions and it was the Gas Council's experience that town gas was much more hazardous, as ready separation of hydrogen occurs.

*Dr Napier* replied that he was not in disagreement with Mr Evans on matters of fact, but he could not agree with some of the inferences drawn. ·For example, conversion to natural gas usage removed the well-known hazards of both hydrogen and carbon monoxide. Considerations must not stop at that stage. Effort must be made to quantify hazards, remove disadvantages, and replace unsuitable or worn-out installations in order to maximize on the quality of natural gas.

He illustrated his call for further assessment by reference to the "explosions" that occur when LNG is spilled on water[23]. Full understanding of the phenomenon, estimation of the extent of electrical charging of the droplets formed, and of the effect of scale were required.

# Natural Gas and the Environment — After Combustion

## By P. F. CORBETT

*(Marketing Services Manager, Gas Division,*
*International Sales Department,*
*British Petroleum Co. Ltd)*

## SUMMARY

Natural gas, as marketed for fuel purposes, is virtually free of adventitious solids and sulphur. In properly designed appliances, the products of its combustion are free from smoke and sulphur oxides. Their contribution to atmospheric pollution, therefore, is minimal as compared with that of the flue gases arising from the combustion of commercially available solid and liquid fuels.

Social and other pressures are establishing environmental control standards aimed at limiting either:

(a) The sulphur content of fuels at the point of combustion and/or

(b) The sulphur oxides content of the resultant flue gases.

These standards will be met only in the cases of solid and liquid fuels by the investment of further capital in flue gas cleaning processes or fuel desulphurization plants. Natural gas, therefore, is in a special class as a premium fuel in pollution control. Its ability fully to discharge its role may be restricted by its limited availability in relation to the total fuel market, and the marketing policies of individual gas distributors who should aim to market the gas at a price which reflects its special advantages.

## INTRODUCTION

A previous paper has dealt with some of the environmental factors associated with the production, storage, and distribution of natural gas: the present paper attempts to expand the study by considering those which might arise during and after its combustion. It is not within the scope of either this paper or the meeting to review atmospheric pollution and other environmental factors extensively, and the method of approach has been

221

briefly to compare and contrast some of the fuels available to the energy markets, and to discuss some of the more obvious pollutants, problems, and potential solutions.

Combustion is a process of oxidation whereby those constituents of a fuel which are "combustible" are encouraged to undergo chemical reactions with oxygen (usually - but not necessarily - diluted with the inert constituents of air) and thereby generate heat. Normally in a chemical reaction, the basic considerations are the purity and yield of (at least) one of the products of the reaction. In this specialized application of oxidation, however, it is the heat which is consequentially generated which is of prime importance. As distinct from the energy derived from their production ( and that still resulting from their being at high temperature) the resultant materials from a combustion reaction are not required. It is these, therefore, which need to be disposed of, and it is their condition and composition which need to be taken into account in an assessment of the effects of their impingement on the environment.

The "value" of a fuel is measured by the amount of heat it is able to generate under ideal and prescribed conditions: the "efficiency" of combustion is the measure of the degree to which results, in practice, approach this theoretical level in terms of quantity of heat generated. Fortunately, optimum heat generation is associated with the most nearly theoretical degree of oxidation. The emergence of unburnt fuel or only partially oxidized combustion products from the combustion zone results in a lower quantity of heat being released than is theoretically possible. Additionally, frequently the condition of the combustion products further impairs the efficiency of other sections of the plant or necessitates extra maintenance or cleaning operations. In the initial design of the plant it is always necessary to consider the fuel likely to be used, its composition, and the efficiency or ease with which it can be burnt. The plant operator and the combustion engineer, therefore, even if not requiring the products of the combustion process, and finally rejecting them to the atmosphere or another part of the environment, are nevertheless concerned (even if indirectly) with the consequences of their initial production of energy.

## COMBUSTION OF FUELS

An elementary sub-division of fuels may be made into those which are (a) solid, (b) liquid, or (c) gaseous, although each group covers a wide range of materials with widely differing properties giving rise to a wide variation in combustion products within each group. Solid fuels include those with a high moisture content such as peat; wood and brown coal; charcoal and coke in their various forms; and bituminous coal which, in most industrial areas, is the predominant solid fuel. Liquid fuels could include an almost limitless range of organic liquids but for the most part are accepted as being those derived from the distillation of crude petroleum oil*. These products vary to the extent that they are distillates or residual and further refined or blended to conform to market or appliance requirements. They vary from the super-refined sulphur-free kerosines required for flueless appliances, such as free standing space-heaters, to the high-viscosity residual fuel oils used in large industrial boilers and furnaces. Gaseous fuels again could cover a wide range of combustible gases (and could include such gases as hydrogen sulphide and acetylene) but for the most part the term is used to cover methane (and other hydrocarbons), hydrogen and carbon monoxide, or mixtures of all or some of these. Originally, these mixtures were derived mostly from the distillation of coal and/or the cracking of oil but the range would now include liquefied petroleum gases and a series of gases generically termed "natural gas".

It is not possible to discuss and compare all these fuels: each makes an individual impact on the environment but not all are equally important. For simplicity's sake it is proposed to consider bituminous coal, residual fuel oil, and natural gas as being the alternative choices available to bulk fuel users in a developed energy market. The first consideration is that of the cleanliness of the fuel itself, i.e the amount of incombustible material likely to be present and carried forward through the combustion process and (if not removed) into the stack gases themselves. The second is the efficiency of combustion, with the possibility (if this is lower than theoretically possible) of unburnt carbon being present in the gases either as discrete particles or conglomerates, or as smoke. The third is the presence of elements which although oxidizable (or partly so) to gaseous compounds and therefore disposable from conventional chimneys, nevertheless  might be contributors to atmospheric pollution. In simple terms it is necessary to give consideration to the ash content of the fuel, the ease with which the fuel can be induced to present the maximum surface area to the air being supplied for combustion, and the proportion of sulphur in the supplied fuel.

## (a) Ash

The ash content of solid fuels is high even after physical separation of non-combustible ores by hand picking and simple washing methods: it may be 10 per cent or more in present fuel preparation and ash disposal problems, both of which are added environmental disturbances. Although steps can be taken to separate most of these solids from the flue gases before discharge to the atmosphere, the installation and operation of grit arrestors and/or electrostatic precipitators is an added cost. Even after 99+ per cent removal the remaining dust still constitutes a pollution problem if disposed of at low level. The ash content of liquid fuels although very much smaller is not nil

*Exception must be made here of a range of liquid fuels - coal tar fuels - derived from the "carbonization" of coal, which are declining in availability consequent upon a decline in this method of manufacturing gas.

and this measurable amount of pollution must be considered in any relative fuel assessment. Even if the solids content of natural gas is measurable at the wellhead the nature of gas distribution (at high pressure and velocity through valves and meters) requires natural gas to be virtually solids-free in its marketed form.

### (b) Completeness of Combustion

Efficient combustion is achieved when the fuel and the theoretical quantity of air are brought intimately into contact. The limiting volume is that of the air and the limiting time for combustion is that of its passage over or through the fuel in the combustion chamber itself. In the case of solid and liquid fuels it is necessary to "prepare" the fuel so that it presents the largest possible surface area to the air and/or to arrange for some proportioning between the air required to project the fuel into the combustion chamber and that required for completing combustion. Although there are cases where solid fuel is burnt on a grate, most modern solid fuel-fired water tube boilers are for the most part fired with pulverized fuel. The solid fuel is ground so that it will mix with (by being entrained in) air, and with proper fuel preparation and design of burner a high degree of combustion efficiency can be achieved. The pulverizing plant, however, is expensive and its operating and maintenance costs are high.

Liquid fuel is prepared by being "atomized" *(i.e.* dispersed into minute droplets) and to achieve this the fuel must be pre-heated to reduce its viscosity and introduced with some air into the combustion appliance via a specially-designed burner which both atomizes the fuel and mixes it with air. From earliest times the best of these have been highly efficient and their development has contributed to the design of burners for solid pulverized fuels. In both cases, however, there remains the possibility that some part of the fuel particle will remain unburnt with consequential effects on the composition of the stack gases. In the case of natural gas the problem of the mixing of the fuel with the combustion air is minimal, since the intermingling of a gas with a gas is the easiest mixture to achieve. Although bad burner design could result in some gas stratification and incomplete combustion, this would normally result only from a deficiency of air. In broad terms the ease (and therefore the potential efficiency) of combustion increases from solid through liquid to gaseous fuels and the likelihood of unburnt fuel contributing to atmospheric pollution decreases in the same order.

### (c) Sulphur

Both solid and liquid fuels contain appreciable amounts of sulphur, the majority of which is oxidized to sulphur oxides and remains present in the flue gases.

At a rough approximation, about half of the sulphur in coal is organic while the rest is present as pyrites distributed non-uniformly throughout the coal measure. Some of this becomes separated by mechanical washing processes, while further amounts could be removed by more refined techniques such as modified flotation (involving close control of the density of the flotation media) and magnetic separation. In all cases, however, a decreased sulphur content is achieved only at the expense of increased wastages of fuel. It i this which puts an economic limit to the lower sulphur content of coals, which, in the case of British coals, averages between 1.5 and 1.8 per cent.

Sulphur in crude petroleum oil is entirely organic and the process of distillation distributes this non-uniformly between the various products and also produces compounds not present in the original crude. A high proportion of the sulphur present in the original crude is retained in the residual fuel oil, the sulphur contents of which vary from below 2 per cent to above 4 per cent depending on the area from which the crude is produced.

Natural gases are not all free from sulphur in their original state, although many are so. Those which contain nil or only a trace of sulphur are normally processed to remove adventitious solids, water, and hydrocarbon liquids only and are then available for distribution to users. When sulphur is present above a defined limit (which varies from country to country but in all cases is very low as compared with the sulphur content of the majority of solid and liquid fuels in the energy market) this must be reduced to conform to a defined standard. The standard is often that enforced by legislation, since in gas markets which were developed in the "manufactured gas" era a high proportion of gas was (and continues to be) burned in flue-less appliances (e.g. cookers, and the smaller types of water heaters) where the emission of toxic gases into inadequately ventilated  kitchens and bathrooms must be guarded against on the grounds of public health and welfare.

In those cases where the amount of sulphur is only marginally above the permitted maximum, the conformity to this standard is an added cost. In some other cases, however, a substanially higher sulphur content makes the natural gas a valuable chemical feedstock for a sulphur extraction process. The best - known example in Europe is the Lacq  field in France, where the wellhead gas has a sufficiently high $H_2S$ content (about 15 per cent) to make its removal and marketing - as elemental sulphur - an economic operation*.

Either because originally the gas had no sulphur content or because the sulphur has had to be removed to conform to legislative standards, it will be seen that for all practical purposes, irrespective of its source and original composition, natural gas as marketed for fuel purposes is a gas mixture predominantly of methane – but also containing small amounts of ethane and other gaseous hydrocarbons – and inert gases. The proportion of each of these constituents will affect the resultant calorific value of the gas mixture

*Some gases in North America, however, have two or three times this amount of $H_2S$ present and are very important sources of sulphur.

which can vary, therefore, from area to area: they are not likely, however, to affect its complete combustion or to add significantly to environmental problems.

## CURRENT SOCIAL AND LEGAL PRESSURE

Quite rightly, most people recognise that an unrestricted and uncontrolled right to inflict inconvenience and/or discomfort on others cannot be sustained. Whereas many years ago most abuses could be corrected only by private legal action, over the last century public law has been developed to cover potential nuisances likely to endanger public health and welfare. Today it is extending its writ to cover a wider range of situations.

The relative ease with which $SO_2$ can be measured and recorded has almost inevitably led to the use of such data at "ground level" (i.e. recording station level) as the most important – and in some cases the only – measure of atmospheric pollution. Although not all authorities agree on the upper limit of $SO_2$ concentration in the atmosphere tolerable on medical grounds, it is argued that sulphur oxides are inherently both toxic and corrosive and therefore some measure of control is socially desirable if not "necessary". Demands in industrial/urban communities to limit environmental pollution have included attempts to limit the effects of sulphur oxides.

In Europe, the U.S.A. and Japan there is an awareness that the environmental factors associated with the combustion of fuels are aspects of social amenity which can and should be controlled by legislation. In most cases, however, the control or proposed control is a somewhat arbitrary one based almost exclusively on the sulphur content of the input fuel without sufficient regard to other factors such as the relationship of pollution to smoke control or the possibility of alternative technical solutions. Proposals which have been made (and discussed later) include the dispersal of untreated flue gases at a relatively high level or the (almost complete) removal of sulphur oxides from the flue gases before their discharge to atmosphere.

In the United Kingdom the Alkali Act has controlled some combustion processes for over 100 years but these are mostly those associated with the smelting or refining of ores or the production of secondary fuels. Some private acts of Parliament subsequently gave some local authorities wider powers to enforce their own standards, but the first major enactment to cover all domestic and industrial uses of fuels was the Clean Air Act, 1956. This prescribed the fuels and appliances which could be used in "control" areas and limited the amount of solid material (in the form of smoke) which could be discharged from any appliance. It had the effect of substantially improving the environment and encouraged the belief that control of the sulphur content of the fuel would enable even further progress to be made by limiting the total level of sulphur dioxide discharged to the atmosphere. Already, the City of London proposes to allow new fuel burning installations on condition that the fuel contains a maximum of 1 per cent sulphur and will require all existing plant to change to fuels of this category over a 15 year period.

Throughout most of Europe the trend is in the same direction. In Belgium, for example, new installations in the five most densely populated towns (Brussels, Antwerp, Liege, Ghent, Charleroi) must burn a fuel with a maximum sulphur content which is related to furnace output. In the lower range this is limited to less than 1 per cent and in no case might the sulphur content of the fuel be greater than 3.8 per cent. Conversion of existing installation must be complete during 1973.

In Italy a maximum of 1 per cent sulphur fuel must be used in installations below a certain size in the six most densely populated industrial towns (Milan, Rome, Venice, Florence, Genoa and Turin). Although no special legislation covers large installations, new ones of the power station type are being "encouraged" to arrange to be supplied with fuels limited to a maximum sulphur content of about 3 per cent.

In France, Paris is the only centre where, as yet, some control is enforced. Certain zones are restricted to 2 per cent maximum sulphur fuel and here again smaller furnaces and boilers have even lower limits imposed upon them.

In Germany, in the large conurbation areas of the Ruhr and elsewhere, some restrictions are placed on fuels. In the most heavily industrialized areas these are restricted to a maximum of 1.8 per cent sulphur, but a further 1 per cent sulphur content is permitted elsewhere where and if ground level concentrations of sulphur dioxide indicate that fuels of this type may be tolerated.

In Austria there is no legislation, but proposals are being discussed for a future limit of 1.5 per cent sulphur. At the moment recording stations for sulphur dioxide are maintained in Vienna, Linz and Graz and at certain levels the municipality may require a change of fuel or the temporary closure of a plant to check the situation. In Switzerland there is talk of imposing a 0.5 per cent sulphur limit on fuels used in domestic installations.

In Scandinavia there has been much recent interest in atmospheric pollution control as one element in the discussions arising from the United Nations Conference on the Human Environment held in Stockholm in June 1972. In Sweden currently no fuels may be burnt with a sulphur content of more than 2.5 per cent and in the three main areas of Stockholm, Gothenburg, and Malmo the limit is 1 per cent maximum. Extensions to the area of the application of this lower limit are proposed and could cover the whole of the country at a maximum of 0.8 per cent except for installations consuming more than about 1000 tons of fuel/year where the limit is relaxed to 2.5 per cent maximum. Although Denmark is so far without legislation for sulphur control, a proposal is under discussion to limit the fuels burnt in Copenhagen to those containing not more than 1 per cent sulphur.

In Holland 31 testing stations for $SO_2$ were installed in 1969 in the highly populated/industrialized area between Rotterdam and the North Sea, and this number is to be increased to 250 by mid-1973. The number in the original control area is to be increased to 110 and the remainder is being distributed throughout the rest of the country. Each station in this "early warning system" is linked via telephone lines to a computer, where $SO_2$, wind velocity, and wind direction are monitored and compared with a mathematical model. If it is calculated that adverse conditions are likely to persist for six hours or more, certain selected industrial fuel users are alerted and instructed to avoid the production of further $SO_2$ until the conditions have been corrected.

## HIGH-LEVEL DISCHARGE OF FLUE GASES

Dispersal of flue gases at a high level has many attractions but is only effectively available to the largest of fuel users, since it relies on providing a sufficiently high chimney to take the gases well above ground level and away from the immediate environment. At this level the large volumes of flue gases involved become diluted with even larger volumes of air and their instaneous effect on atmospheric pollution is correspondingly diminished. Chimneys in the early Industrial Revolution had a two-fold purpose: one was to disperse relatively large volumes of smoky flue gases above the roof-tops of the neighbouring houses (this was largely ineffective as most early industrial communities grew up in valleys along natural water supplies and easy communication routes), the other was to induce sufficient air through the fuel bed (usually coal) to produce some degree of combustion. This also was only partly effective and a combination of smoke and sulphur oxides at low level gave rise to fogs and generally unacceptable levels of dirt and pollution. There was progressively gradual improvement, mostly associated with increasing the efficiency of fuel burning, and the installation of induced and (in some cases) forced draught fans made the control of combustion air more scientific. As boiler and furnace plant became larger chimneys were needed to accommodate combustion products from a number of integrated appliances; this led to considerable development in chimney design and more understanding of the effects of chimney height and weather conditions on gas plume rise and product dispersal.

It was, however, the electricity industry in the post-war years which contributed most to the understanding of this type of anti- (ground level) pollution technique. As power stations became larger and quantities of fuel of the order of a quarter to half a million tons/year were consumed at one site, it became necessary to build high chimneys so that, with the natural thermal lift of the gases and the movement of air at levels of 300-400 ft above ground level, the sulphur oxides would be diluted and less noxious when and if finding their way to ground level.

So successful has been this policy that in spite of increases in the quantities of fuel used by industry, the ground level concentrations of $SO_2$ have in fact declined over the last 10-15 years. Domestic fuel usage has also changed in character in many areas as a result of the Clean Air Act. More gas and electricity is used for single room heating than hitherto, and gas and oil-fired central heating, and coke stoves, are replacing the "traditional" (open) coal fire.

These trends would both tend to lower the ground level concentration of $SO_2$, but without a policy of "dispersion" it is likely that the local effects of large power stations would have more than "compensated" for this improvement.

Over the last 20 years the $SO_2$ emission from electricity generating stations (based on fuel consumption and sulphur content) has increased from about 1 to 2 million tons/year derived from coal and from about 10,000 to over 1.5 million tons derived from oil. Overall, the total emission of $SO_2$ from all sources and appliances has gone up by about 20 per cent whereas the ground level concentration has been virtually halved.

On the national scale, therefore, UK industry has demonstrably proved the efficiency of this policy. Internationally, however, there is criticism as countries windward of the UK (Scandinavia) have suggested that inconsistencies in their calculated sulphur oxides output and ground level concentrations of $SO_2$ can be explained by adventitious "import" from the UK.

## DEMAND FOR LOW-SULPHUR FUELS

Pressures to restrict the sulphur content of fuels will put these pressures directly on to their producers and suppliers. Legislative (rather than persuasive) action will accelerate the already observable trends.

It has been mentioned that the physical separation required for further refinement of coal is difficult both to achieve and justify. Unless coal is gasified, it is unlikely to lend itself to chemical desulphurization processes. In view of these technical difficulties and the declining share that coal has of the energy market, it is unlikely that legislation will require the reduction of sulphur in coal.

There has always been a demand for low-sulphur fuel oils and hitherto these have been met. Processes where the products of combustion are in contact with the material being processed (e.g. in glass making, steel producing, direct drying processes, etc.) could only be supplied by oil, in competition with other fuels, when special arrangements were made to make low-sulphur fuel oil available. This was achieved by a conscious allocation of fuels derived from a low-sulphur containing crude, and needed special and separate storage and distribution facilities from those of "normal" sulphur content. Demands for this selective treatment are still being met by the allocation or re-allocation of supplies, even though today, in many cases, the fuel is not being used totally in these former "premium" markets.

In many areas (France, Sweden, Germany, the Netherlands) the extra costs incurred in this segregation are still reflected in the price that consumers pay for a low sulphur-containing fuel.

It is unlikely that very much further re-allocation is possible. Low sulphur-containing crudes are not available in such quantities as to cater fully for the present-day potential demand. Even if further sources became available from newly-discovered fields, it is probable that any "home" government would insist on priority claims to such sources.

The situation can be met in a number of ways and those likely to be developed involve coking and gasification to sulphur-free or low-sulphur fuels. Continuing work is also reported on desulphurization during combustion itself using slightly basic additives and a fluidized-bed technique, but such processes are unlikely to be available on the large scale to meet the almost instantaneous demands that are foreseen.

The main lines of immediate application are the desulphurization of fuel oil itself. Distillates have been further refined for some time, but only within the last years has there been the necessity to produce lower sulphur contents for residual oils. The methods used (and which are immediately available for further application) have involved hydro-desulphurization, where a fixed-bed catalytic hydrocracking process using high-pressure hydrogen converts the sulphur in high molecular weight compounds to hydrogen sulphide which can then be removed. A wide variety of cost estimates is available and a simple yardstick is approximately £2 capital investment/one ton/year throughput. The overall cost is very significantly affected by the rate charged for capital but even at the lowest the cost of the process is considerably more than that involved in the re-allocation of restricted low sulphur crude supplies.

## REMOVAL OF SULPHUR OXIDES FROM FLUE GASES

If it is considered that, irrespective of the point of discharge, the emission of $SO_2$ from chimneys must be controlled, an alternative to removing part or all of the sulphur from the fuel before combustion is to remove the resultant sulphur oxides from the gases themselves, after combustion but before their discharge to atmosphere. This puts the onus of meeting control regulations or requirements on to the fuel user directly and involves *him* in the acquisition, operation, and maintenance of some suitably-designed technical plant and process. Quite clearly the capital outlay in relation to the quantity of fuel used and technical control required to operate and maintain the selected process, limit the possibility of such applications to the largest fuel users only. In fact, it is generally accepted that the only fuel burning plants likely to overcome the very many tests of technical and economic feasibility are concentrations of steam-raising plants at electricity generating stations or plants of comparable size. With this application in mind, it is perhaps useful to consider some of the simple calculations which arise.

Assuming that a 4000-MW station is burning residual fuel oil, with a sulphur content of about 2.5 per cent at the rate of about 25,000 tons/day, then about 20,000 million ($20 \times 10^9$) cu ft of hot flue gases will be emitted per day and require treatment for the (almost complete) removal of the sulphur oxides contained in them. The gases themselves, however, are mainly nitrogen and carbon dioxide, the sulphur oxide content being only about 0.2 per cent of the total volume.

It will be seen, therefore, that the volume of gases required to be treated is out of all proportion to that of the volume of gas required to be removed. The handling of such large volumes requires the building of a complex recovery plant which could take up more space than the steam and electricity generators of the originally designed plant. Furthermore, the removal of the direct effects of gaseous effluent is not without some consequential side-effects. Further attention must be directed to the disposal of effluents (or their re-treatment for re-cycling), the effects of saturated water vapour and a lower temperature on the treated flue gases, and the disposal of the main "product" on the market. Not every process being developed has elemental sulphur as its by-product but, assuming this to be the case for the hypothetical 4000-MW station discussed above, it will be necessary to dispose of about 600 tons of sulphur each day. The operations of conveying, storing and packaging this material constitute further additions to the plant facilities, and new environmental problems.

There are about 50 processes available to choose from but not all of these are fully developed and only a few have been tested. Their number is a measure of the possible combinations of a variety of factors on which a process is dependent. These include the type and sulphur content of the fuel, location of plant, type of grit arrestment, boiler or combustion system, and the process end-product required or which can be tolerated.

There are many considerations in the evaluation of a process which lead to the relative assessment of one process with respect to another, but the efficiency with which sulphur oxides can be removed; the capital, operating, and maintenance costs (i.e. the economics of the process); and the character of the resultant by- or waste-product (i.e. its composition, purity, and potential marketing and/or disposal) are the most important.

Some of the processes are claimed to be "economic" in the specific circumstances of their application but, even if accepted as such for prescribed conditions, are not necessarily so under different ones. All projections of the economics of flue gas washing appear generously optimistic and in too few cases have been corrected for inflationary tendencies and currency de- or re-valuations. Perhaps the most easily overlooked factor which affects the economic viability of a flue gas treatment project is the load factor of the plant to which it is "tied". In the case of a power station this can vary significantly over its operational life. In the first five years or so a newly-commissioned station can expect to be among the most efficient and therefore one of those selected to operate at "base-load"; this may result in a load factor over this period of about 90 per cent. But as newer stations supersede it technically and economically, its load factor will decline. At the end of its operational life, it could be on stream for only a few hours each day and eventually working for short periods only on the coldest days of the winter months. In these circumstances the *average* load factor over the whole plant life could be around 50 per cent or less. Even though the flue gas desulphurization process may be able to cope with this technically, it is doubtful if it could be justified economically.

The choice of end-product will clearly be influenced by market considerations; it may also be dependent on the location of the site where one product may have some special requirement. The market value of the by-product presents special problems as the value (if any) assumed for income from this source affects the economics of the process. New sources of material affect the balance of supply and demand and could affect the original assumed market price. As an example of this possibility it has been calculated that if all the sulphur contained in the flue gases from power plants were extracted and marketed as elemental sulphur, the world availability of sulphur would be doubled. It is doubtful if quantities of this order (or even smaller) could be absorbed on the market without a serious downward price movement and its consequential effect on the whole economics of sulphur production, including that from flue gas desulphurization processes themselves.

Site location, as well as influencing choice of end-product, may also seriously affect the process economics: it is a factor over which there is little control. Whereas in earlier times electricity generating stations tended to be built within a defined and individual franchise supply area which was usually urban, developments in transmission and the economic size of stations and the quantities of fuel required at these have led, over the last 20 years or so, to power stations being sited on coal measures or on estuaries or coasts where they can receive oil tankers. Many sites, such as those adjacent to oil refineries and/or chemical plants, may not be so affected, but an unfavourable location for a gas treatment plant may add an element to the distribution costs of an end-product such as to render uneconomic one which under other conditions might have some return value.

## THE ADVANTAGES OF NATURAL GAS

This brief essay has been concerned with establishing some advantages for natural gas in the field of environmental control. Whereas the products of combustion of soil and liquid fuels cannot be entirely freed from solid particulate matter nor (without considerable investment) be freed even partially from sulphur oxides, flue gases arising from the combustion of commercially-distributed natural gas may be considered to be free of both of these potential pollutants. These important differences confer considerable advantages on fuel users who choose a gaseous form of energy.

In the purely domestic market there are many unquantifiable factors affecting the choice of fuel: ease of supply; social habit; pressure of advertising, friends, and neighbours; local and national proscription and prescription; the relative values placed on convenience and maintenance, as well as the initial cost of the appliance and the continuing cost of the fuel – all outside the scope of this summary. In the industrial market however, under unrestricted conditions, the user's choice is governed by a calculated assessment of the inter-connection between capital investment, maintenance costs, and fuel costs for each of the fuels he is able to procure on a continuing basis. If his choice is modified by either a legal or a moral obligation to consider atmospheric pollution, some additional consideration will apply and effect each of these three main areas of decision.

In the original design of the plant (i.e. initial capital investment), the choice of natural gas will avoid considerations of fuel storage, tall chimneys, grit arrestors, and flue gas washing, and the selection of fuel preparation and combustion equipment will be less complicated and costly than in the cases of solid and liquid fuels.

Maintenance on much of this equipment will be avoided, and on that remaining will be considerably reduced. Oxides of sulphur (particularly sulphur trioxide), either with or without solid materials also present, could cause deposition, corrosion, and maintenance problems in large boiler plant and could limit the degree to which heat may be extracted from the flue gases. It is possible, therefore, that in boiler plant specially designed for burning natural gas a higher degree of overall fuel efficiency could be achieved.

However, it would be prudent to avoid too many favourable claims in these areas of decision-making. In many cases the discontinuous availability of natural gas may limit the unrestricted freedom of a fuel user to make all these decisions for himself. Large single points of fuel consumption, where specialized staff is available to make the necessary and consequential changes, are, by their very nature, ideal ones to be selected for the restriction, or even the curtailment of gas supplies when demands from a less readily controlled domestic market are temporarily in excess of maximum supply. Such interruptible gas supplies may be voluntarily accepted under special conditions of contract between the user and gas supplier or, in the extreme, may be legally enforced. In either event, the effect would be for the industrial consumer to have a plant equipped to burn both gaseous and liquid fuel when many of the special advantages in favour of an initial design for gas only are lost at the design and investment stage.

But the savings on maintenance costs and the ability fully to comply with a very low level of sulphur oxide emission will be present in proportion to the time that the plant is able to secure uninterrupted gas at a load factor which conforms to its own pattern of supply and demand. In cases where the industrial user (whether electricity or general) can arrange his own direct supplies of gas, these problems are diminished.

There are insufficient reserves of natural gas to supply all consumers demanding special low sulphur fuels and some considered policy of allocation might ultimately be reached. In all cases, gas will be more valuable than the fuels with which it is in competition, and the costs of these will increase to the extent that desulphurization is necessary and special distribution arrangments must be made.

## DISCUSSION

Discussion was opened by *Dr A. R. Khan* (Gas Development Corpn) who, after thanking the organizers for inviting him to attend, commended Mr Corbett on "a brief but meaningful survey of natural gas and its advantageous position relative to the environment". The U.N. Conference now meeting in Stockholm indicated man's logical concern over his environment today.

While natural gas was the cleanest of the fossil fuels and alleviated the problem of particulate matter and sulphur oxides, the combustion of natural gas, as with other fossil fuels, generated oxides of nitrogen at high temperatures, and their concentration increased with increasing combustion intensity. Several U.S. studies had shown that $NO_x$ concentrations range from 50 ppm in small heaters to as much as 1500 ppm in large power plant boilers. In a recent study the Environment Protection Agency of the U.S. Government reported that the combustion of different fuels produced roughly comparable $NO_x$ emissions per unit of energy, with gas accounting for the lowest amount compared with coal and oil.

The problem of $NO_x$ was different from that of other pollutants, as $NO_x$ formed whether or not nitrogen was present in the fuel. CO and other hydrocarbon pollutants formed because of incomplete combustion. $NO_x$ by contrast formed because of complete combustion. In fact, most of the "improvements" in industrial combustion systems in recent years had led to greater emissions of $NO_x$. The increase in combustion intensity had led to very high concentrations of $NO_x$ resulting from the complex chemistry by which $NO_x$ were formed.

The technology for control of oxides of nitrogen was based on matching the rate of heat generation by combustion to the heat requirements of the load. For $NO_x$ control the burner should be designed to give a lower rate of combustion than for conventional burners, so the gases never reached the very hot condition. IGT had shown that this technology could be successfully applied, and the gas industry was continuing to sponsor research to further improve its ability to decrease this type of pollution.

*Dr M. J. Stacey* (British Oxygen Co. Ltd) asked what was the pollutant effect of $NO_x$, and therefore the concern over it.

*Dr Khan* indicated that the concern resulted from the fact that it had been established that $NO_x$ were the catalysts which created the smog hanging over Los Angeles. It was essential for the gas industry to act now to forestall public outcry.

*Mr Corbett* pointed out that it was not possible to look at the effect from one chimney or exhaust since the conditions in which $NO_x$ acted as a catalyst resulted from the effect of many different combustion gases from many sources interacting in a small area and believed to be promoted by bright sunlight.

*J. H. Dick* (Gas Council) said that the Gas Council (through the Midland Research Station) was taking the closest interest in the $NO_x$ problem. Scientists working on this problem were exchanging visits with IGT. In the U.K. the problem was not the same as for Los Angeles, where there was linkage with hydrocarbons in a limited basin under special climatic conditions.

Mr Dick mentioned the paper presented to the Institution of Gas Engineers in May by Professor Alice Garnett, of Sheffield University, who showed that the drop in $SO_2$ concentration measurements recorded in that city could be attributed to the changeover to natural gas.

*J. A. Field* (NCB (Exploration) Ltd) thought that Mr Corbett seemed to dispose of the carbonizing industry in a few words, but the NCB was still able to sell on long-term contract all the coke oven gas available.

*Mr Corbett* assured Mr Field that it had never been his intention to give the impression that the coke industry had no future, but for the purposes of his brief review he had compared natural gas with coal and oil, which were universally available alternatives.

*D. Elgin* (Scottish Gas Board) considered that SNG, if one took into account its combustion characteristics and transportability, provided a means of turning fuels that could be more harmful in the environment into less harmful fuels.

*Mr Corbett* pointed out that the conversion of primary to secondary energy leads to a loss of energy and in a fuel short position this might not be so attractive. But certainly the advantages mentioned by Mr Elgin could stimulate the production of SNG.

# Natural Gas in the Future Energy Picture

By M. W. H. PEEBLES *(Shell International Gas Ltd)*
and M. D. J. GELLARD *(Shell International Petroleum Co. Ltd)*

---

ABBREVIATIONS

| | | |
|---|---|---|
| mrd m$^3$ | = | milliard ($10^9$) cubic metres |
| t.c.e. | = | metric tonne of coal equivalent |
| b/d o.e. | = | barrels per day of oil equivalent |
| MMBtu | = | million ($10^6$) Btu |
| MMcf/day | = | millions of cubic feet per day |
| TCF | = | trillions ($10^{12}$) cubic feet |
| GNP | = | Gross National Product |

APPROXIMATE CONVERSION EQUIVALENTS

| | | |
|---|---|---|
| 1 mrd m$^3$/year | = | 100 MMcf/day or 0.04 TCF/year |
| | = | 17,600 b/d o.e. or 900,000 tons fuel oil/year |
| | = | 1.4 million t.c.e/year |
| | = | 11,000 million kWh/year |
| 1 million t.c.e. | = | 0.73 mrd m$^3$ |
| | = | 27,300 million ft$^3$ |
| | = | 8,000 million kWh |
| | = | 4.7 million barrels of oil equivalent |
| | = | 65 tons of Uranium oxide (notional) |
| 1 TCF | = | 27 mrd m$^3$ |
| 10 therms | = | 1 million Btu |
| US ¢10 per MMBtu natural gas | = | US ¢63 per barrel fuel oil |
| | = | 0.92 old pence per therm natural gas |
| | = | 0.38 new pence per therm natural gas |
| US $1 per barrel fuel oil | = | US ¢16 per MMBtu natural |

237

Note: For the above conversion equivalents natural gas is assumed
to have a calorific value of about 1000 Btu per ft$^3$ or 9400
kcal per m$^3$. The price conversions are based on £1 = US
$2.60, i.e. as at February 1972.

## THE CURRENT WORLD ENERGY SCENE

The current role and future development of natural gas cannot be viewed
in isolation from the overall context of the world energy scene in which it
moves. The basic trends underlying world energy supply and demand are
inherent in any understanding of the present and future contribution to be
expected from natural gas.

Natural gas is included with petroleum fuels, solid fuels, hydro, and nuclear
electricity amongst the so-called commercial forms of primary energy. The
non-commercial forms, such as wood and dung, have lost much of their former
significance, although in some countries, for example in India, they are still
important. However, hereafter in this paper the term "primary energy" exclude
the non-commercial forms.

The location and availability of the world's supplies of energy obviously
have significance in determining the relative economics of the different fuels
and the subsequent extent and form of their usage within individual countries
and regions. For this reason, although reserves of primary energy are by their
nature notoriously difficult to estimate with any accuracy, it is nevertheless
necessary to look at the location and extent of the major reserves of energy
in the world.

Fig. 1 shows the main locations and gives an indication of the general order
of magnitude. Proven reserves of oil, which can be regarded as a rough measure
of the oil industry's working stock of oil in the ground, are put at approximately 116 billion (10$^9$ million) tons of coal equivalent at the end of 1970,
with ultimate recoverable reserves possibly three or four times as much. In
addition, new discoveries are continually being made on land and on the
Continental shelves, and perhaps also eventually in the deep oceans, that may
alter the distribution pattern quite considerably. However, the already
dominant position of the Middle East, with over 60 per cent of world oil
reserves, is likely to continue through to the mid-1980s.

Oil other than in underground reservoirs, such as in oil shales or in tar
sands, is still awaiting the right economic climate or a technical breakthrough
for large-scale commercial exploitation. But these enormous reserves of oil
do represent a substantial fall-back reservoir of energy for the world which
has already been located, albeit mainly on the American continent, and
which it is technically possible to utilize.

The world's present total proven natural gas reserves are of the order of
65 billion tons of coal 'equivalent (or, say, about 49,000 mrd m$^3$), representing
over a third of total conventional hydrocarbon reserves. Fig. 1 shows the
quantitative distribution of these gas reserves by main geographical regions.

# World Energy Reserves

# Shares of Market Demand 1969/70

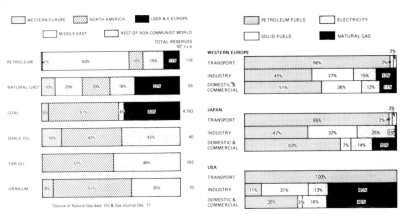

Fig. 1

Fig. 2

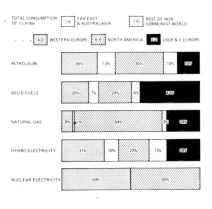

**Energy
Consumption by
Region 1970**

Fig. 3

It also shows the location and rough order of magnitude of the world's coal reserves, both hard and brown coal. Other estimates for coal reserves range up to as much as four times the size of the ones shown here, and in comparing the size of these reserves with those of oil and gas, the extent to which economic recoverability is really possible must be considered. The enormous strength of the North American and the Communist area coal reserves is, however, clear, as is the reversal of strengths between Europe and the Middle East when compared with the position of oil reserves.

The large share of uranium oxide reserves shown for North America may represent merely the results of more intensive surveys and available information rather than a true picture, and possible additional world reserves of uranium oxide have been estimated around two and a half times higher.

From Fig. 1 it is clear that coal in terms of sheer volume available appears to be the most important single source of energy. However, in considering the interplay of forces leading to demand for a particular form of energy it is necessary to look at the demand for energy in its final form as required by the consumer. Fig. 2 shows current demand for energy in the main final market sectors of leading energy-consuming regions.

The growth of total energy demand is a close reflection of the economic progress of a country. But the relationship between economic activity and energy consumption is by no means simple. Much depends on the relative importance of the industrial, agricultural, transportation, and other sectors in the economic structure of a particular region or country. The pattern of the energy sectors is of importance in determining the extent and speed with which a particular form of primary energy such as natural gas develops. A particular fuel may be price-competitive but may have access only to certain market sectors. For example, the road, sea, and air transport markets are, and are likely to remain over the next two decades, to all intents and purposes virtually the exclusive preserve of oil. The iron and steel industry for at least a decade will continue to require massive quantities of coking coal despite the growth of new processes thus providing the coal industry with a continued high volume outlet for its best quality material.

Relative efficiency in end-use of the different forms of energy available to the consumer is also an important factor in economic assessment of the choice available. The efficiency in end-use of electricity is higher than a fuel such as natural gas, but electricity's actual, much lower, efficiency is concealed within the transformation process. Average efficiency at which thermal electricity is generated is currently around 32 per cent in Western Europe which compares with direct fuel uses of the various hydrocarbons of 50-90 per cent.

Fig. 3 shows the distribution of energy consumption by main consuming regions in 1970. The great industrialized regions of Western Europe, North America, Japan, and U.S.S.R. accounted for approximately 80 per cent of world energy consumption in that year.

It is instructive to compare the patterns of primary energy demand in the main industrialized regions. North America, which has large and highly developed gas resources, has the highest proportion of natural gas and is the world's largest gas market. To date it has also been comparatively self-sufficient in its energy requirements, as opposed to Western Europe and Japan. In the U.S.S.R., where ample proven gas reserves are in-place but development was begun later than in the U.S.A., a little over 20 per cent of total energy requirements are supplied by natural gas. Natural gas demand in Western Europe is based on the comparatively recent development of indigenous reserves and its share of total energy is, as yet, comparatively small, whilst petroleum fuels, most of which are imported, represent the largest single source of energy. Japan, with very limited indigenous energy, is even more dominated by petroleum fuels with natural gas playing a minor role in 1970.

Hydro-electricity contributes only a little over 6 per cent to the world's demand for energy and nuclear power, and, with a less than 1 per cent share, is as yet comparatively insignificant in its contribution.

## DEVELOPMENTS THAT COULD AFFECT THE CURRENT ENERGY SCENE

In the course of the 1960s world energy consumption (outside the Communist areas) rose at an annual average growth rate of 5.2 per cent. In comparison, the growth of both oil and natural gas has been at over 7.0 per cent. One of the major reasons for the above-average growth experienced by oil has been its substitution for solid fuels in the industrial, space heating, and electricity generation markets. Natural gas has grown at the expense of oil as well as coal and the extent to which this substitution process, as opposed to the normal growth associated with economic advance will continue depends basically on the availability, price levels, and quality of the competing fuels.

The low oil prices of the last decade or so that have led to such rapid growth by oil are becoming a thing of the past. The recent price increases for crude oil, and the prospect of further regular increases to come, are bound to have an effect on the competitive position of oil. These higher oil prices, together with the increasing dependence on the Middle East and African reserves, will stimulate the development of alternative energy resources, and probably lead to full-scale reappraisals by Governments of their energy policies.

The outlook for coal is being re-assessed with a growing realization of the underlying strength of coal reserves. A large expansion of coal demand is forecast for both the U.S.A. and the U.S.S.R. mainly in the iron and steel industries and power generation markets. New markets are also likely to develop in the U.S.A. and perhaps elsewhere, such as hydrogenation of coal to make up the growing shortage of indigenous oil production and gasification of coal to offset the declining U.S. resources of natural gas. The U.S.A. is already the world's largest exporter of coal and, with its enormous resources, is, in theory, able to supply Western Europe and Japan with much of their growing energy needs. U.S. steam coal could be laid down in North-West European ports at price levels well below the cost of mining most indigenous West European coal. However, various long-term factors which tend to lead to increases in U.S. coal prices, and the increasing awareness within the U.S.A. of a need to husband its own resources, may restrict any expansion of exports on a massive scale.

In electricity generation and to the extent that if affects the relative economics of electricity in other markets, nuclear power is an important competitor. There have been, however, many serious delays and difficulties and over-optimism with regard to costs, which have led to a slow and disappointing rate of development. Nuclear's contribution to total energy demand is still very small, a little over 1 per cent in Western Europe and under 1 per cent in North America, and its future appears apparently to be still just over the horizon. For the base load generation of electricity, break-even prices quoted currently in the U.S.A. are around $2-$4/brl of fuel oil, and at these prices nuclear must eventually become dominant in this field but there will always be a need for other more flexible, less capital-intensive power stations fired by oil, gas, or coal to cover the intermediate and peak loads.

Although some 90-95 per cent of the world's hydro-electric potential awaits development, most of the remaining significant sites are in remote and capital-scarce areas of Africa and South America. This limitation on available sites will always restrict hydro to a minor source of energy.

Growing public and Government concern over environmental problems leading to restrictions on nuclear plant siting, the control of emission levels and of sulphur content, etc., are also influencing the competitive relationship between the energy sources. Using natural gas to burn under power station boilers may be a waste of its inherent premium qualities, but it is a very attractive proposition to countries like Japan, where the permitted sulphur content of fuel oil in overpopulated areas will be 0.77 per cent by 1975.

The development of a storage battery with adequate capacity and power density for general use in transportation, if successful, could eventually lead to replacement of oil in the transport sector. Natural gas in liquefied form (LNG) may also have a limited part to play in the automotive market if exhaust emissions from gasoline engines cannot be reduced to acceptable limits, and longer term LNG may be used as a fuel for supersonic aircraft.

In electricity generation, the tendency to maximize unit size in power plants will be encouraged by successful developments in high voltage transmission. The use of MHD power production, generating power at up to 60 per cent efficiency, and of fuel cells with a potential efficiency of over 60 per cent will also conserve energy and at the same time provide new competitive stimuli. Increased efficiency will tend to be counterbalanced by the continuing and growing desire for comfort and convenience as disposable incomes increase and it is here in such premium markets, that natural gas, by its inherent qualities, has a particular advantage.

## NATURAL GAS SUPPLY/DEMAND CONSIDERATIONS AND ITS EVOLVING ROLE IN THE MARKET

As stated above, the U.S.A. is the world's largest gas market. The history and characteristics of the American gas industry are so well documented that we will not engage in yet another exposition. Suffice it to say that in this vast energy-consuming country, natural gas today meets about one-third of total primary energy requirements, reflecting the availability of abundant supplies in the past at artificially low prices. The cumulative inhibiting effect of the FPC's regulatory policies is, however, now being felt and over the last three years or so the discovery of new reserves has not kept pace with the continual rise in consumption, with the result that the reserves : annual production ratio has declined from over 16 : 1 in 1966 to under 13 : 1 in 1970. Most observers of the U.S. energy scene accept that the country is entering a critical gas supply situation to the extent that even after allowing for new indigenous discoveries, including Alaskan North Slope reserves, imports from Canada and LNG imports from foreign sources backed up by the development of new processes to make synthetic natural gas from coal and/or oil, the U.S.A. will have a massive supply deficit by the 1980s if gas is assumed to retain its current one-third share of total energy consumption.

Obviously in this situation all indigenous resources will be developed to the maximum extent possible while hopefully large-scale imports will be permitted, even encouraged, by the regulatory authorities. Nevertheless, even under the most favourable supply predictions a change in the pattern of demand in the market place is essential if the best possible use is to be made of this high-quality premium fuel. It is too early to say with any precision how the gas market in the U.S.A. will adjust over the next ten years or so but it seems probable that the residential sector will receive priority and will continue to expand in contrast with some low grade under-boiler applications, which are likely to be progressively converted to other more appropriate fuels. The market value of gas is already moving upwards and will continue to rise, possibly to the point where a distributor in the New York area currently paying around U.S. ¢45-50/MMBtu for indigenous supplies can within ten years expect to pay from about ¢70 to ¢130/MMBtu for new base load supplies, depending upon their source.

The future for the American gas industry can be summed up by saying that as supply constraints bear down on the market to an ever-increasing extent, the resulting greater concentration on high-price premium outlets should provide an increasingly favourable climate for imports and for new gas-making processes. Nevertheless, quantitatively gas will probably progressively lose market share by up to ten percentage points over the next ten to fifteen years.

Turning to the world's second largest gas market, the U.S.S.R., here one sees a completely different picture. As previously mentioned, the development of gas came later than in North America, but already it supplies about 20 per cent of the U.S.S.R.'s total energy requirements. The U.S.S.R. has ample proven reserves – currently estimated to be over 15,000 mrd m$^3$ – and substantial additions are announced each year. Ultimate reserves are claimed by the Soviets to be as high as 70-80,000 mrd m$^3$ – approaching twice the present total proven gas reserves for the whole world. How much of these expectations can be proved up and developed is an open question with immense technical and economic problems to be overcome before these expectations, which lie largely in the remote Siberian regions, can be brought to market. However, if gas is given the requisite development priority, sufficient reserves undoubtedly exist to supply all foreseeable internal requirements whilst still leaving ample capacity for very substantial exports to be undertaken. Already, export deals have been made with Austria, Italy, West Germany, and France and follow-up contracts with these and other countries such as Finland, Sweden, Japan, and even, perhaps the U.S.A. can be expected. Soviet gas is undoubtedly on the move.

The third main gas market is Western Europe. Ignoring the U.K. for the present, the main indigenous reserves, all non-associated gas, are located in the Netherlands, West Germany, France, and Italy, supplemented by recent discoveries of both associated and non-associated gas in Norwegian and Danish waters; see Figs. 5 and 6. The story of the development of the Groningen reserv the Western World's largest non-associated gas field, is well known. Suffice it to say here that by the late 1970s Groningen gas will be supplying about one half of all energy used in the Netherlands, with large volumes being exported to Belgium, West Germany, France and Italy. At plateau production Groningen will be producing more than twice the maximum estimated availability from the U.K.'s North Sea present known reserves.

While the Netherlands is well provided for, other countries, such as West Germany, France, Italy and Austria are increasingly turning to imports to supplement their limited and already fully committed indigenous reserves; Belgium already relies exclusively on supples from the Netherlands. If gas is to expand beyond, or even maintain, the shares of total energy demand which it will have reached in these countries by the late 1970s, further projects to supplement existing/planned purchases of Dutch, Soviet, and North African gas will be essential in the absence of any major new indigenous discoveries.

The last noteworthy gas market is Japan. It is the world's most rapidly developing energy market but with very limited indigenous energy resources. The  potential demand for a clean non-polluting fuel such as natural gas is virtually unlimited. Already 1.4 mrd m$^3$/year of LNG is imported from Alaska and a further 7 mrd m$^3$/year will be added from Brunei, with initial supplies commencing end 1972. Other import possibilities that could develop include LNG from Sarawak, Australia, and the Middle East, also Soviet gas either as LNG or by undersea pipeline. Supplies may also be forthcoming from the

# U.K.–Inland Energy
# Consumption and Forecast 1950–1976

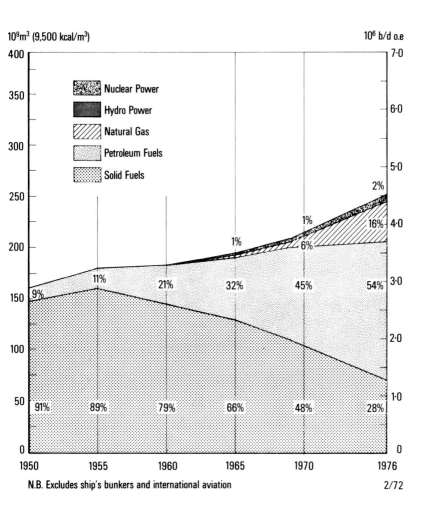

N.B. Excludes ship's bunkers and international aviation

2/72

Fig. 4

extensive offshore Japanese concessions now being explored. Like most new markets, gas will need to penetrate a whole range of applications from domestic outlets to the large industrial consumer in order that economies of scale, high load factor market mix, and the right blend of market values can be realized to support imports which may well be at c.i.f. Japan prices around ¢100/MMBt or more by the late 1970s and early 1980s.

To sum up thus far, apart from the U.S.S.R., where reserves appear to be abundant, the future role for natural gas in the main energy-consuming markets of the U.S.A., Japan and Europe (with the single exception of the Netherlands) is quite clearly more likely to be affected by limitations of supply rather than of demand. As the supply factors become more stringent, so will market development need to become more selective. Market values will therefore rise not only as the result of general inflationary pressures and increasing price levels for energy as a whole, but also in relation to competing fuels as a reflection of supply availability and the increasing concern for the environment, which places an additional premium value on a high-quality clean energy form such as natural gas.

## NEW SOURCES AND DEVELOPMENTS AFFECTING GAS SUPPLIES

In the foregoing section we have described the pull and push of gas supply and demand that is emerging. What then of future development prospects for gas?

First, all conventional indigenous reserves, including new discoveries, located near large energy markets will be developed to the maximum possible extent, constrained only by any adverse technical, political, or economic factors which might exist in certain individual cases. To this can be added the development of undersea resources from ever-increasing water depths as exploration and production techniques improve, provided that suitable financial incentives by way of high market prices for the resultant gas are forthcoming. Similar considerations apply to the exploitation of deeply located onshore gas, nuclear stimulation of tight reservoirs, etc.

Secondly, gas reserves located in areas remote from the main markets will be increasingly exploited. Here the development of LNG technology offers the opportunity to produce, liquefy, and transport over several thousands of miles gas which otherwise would have either to be flared or left idle in the ground. Not all such reserves are technically or economically suitable for development. But certainly the rapid strides made in the last few years in the LNG field are widening the catchment area. By the mid-1980s, it is not unreasonable to predict that, given a favourable economic and political climate, the world's international LNG trade could be well in excess of 200 mrd m$^3$/year (compared with 7 mrd m$^3$/year in 1971), directed principally at the U.S., European and Japanese markets. A business of this dimension could require a total cumulative capital investment of perhaps some £15,000 million for the construction of liquefaction plants, LNG tankers, and the necessary gas production facilities and reception/re-gasification terminals. The implementation of such a massive programme over the next 15 years or so presents an exciting challenge to the gas industry. Provided that the business is sufficiently attractive financially and there is good co-operation and mutual understanding between all the various parties involved, then in a gas-hungry world there is no reason why an international LNG trade of this order of magnitude should not develop.

Thirdly, are the prospects of new processes to produce synthetic natural gas from coal and/or oil. For these to be viable it is necessary, *inter alia,* that the end-product is compatible with existing natural gas supplies and that the process plants themselves are environmentally acceptable. A considerable research effort is now being mounted in the U.S.A. into various coal gasification processes which could be in commercial operation by around 1980. One responsible American estimate is that by the mid-1980s 25 such plants, each producing about 2.5 mrd m$^3$/year, could be in operation requiring some 150 million tons/year of coal. While the U.S.A. has vast reserves of coal and could financially support such a programme, it remains to be seen whether the public and the authorities will accept the environmental consequences involved in producing the coal required. Oil gasification processes will also be developed, but on a more restricted basis, due to the limited availability of suitable light distillate feedstocks. Both gas from coal and oil will not be cheap and could well exceed ¢100/MMBtu ex-plant.

Let us now come nearer to home and take a more detailed look at the U.K. situation.

## THE U.K. GAS SUPPLY/DEMAND SITUATION

The very considerable exploration effort which has been mounted in the North Sea has resulted in the discovery of significant gas reserves in U.K. waters. More precisely, the reserves are sited in five main offshore fields supplemented by a rather greater number of smaller deposits, including one on-land at Lockton in Yorkshire, see Fig. 5. There is undoubtedly more gas to be found under the North Sea, but the fields are likely to be smaller and more difficult and costly to exploit than the initial discoveries. Some may indeed never figure as recoverable reserves at all, given that the effect of the low prices so far paid to the producers of existing reserves has been to reduce the incentive to them to continue exploring, at least for gas, in the remaining less-promising areas.

Total known recoverable reserves under the U.K. continental shelf now amount to some 800 to 900 mrd m$^3$, to which may be added prospective reserves of, say, a further 200 to 300 mrd m$^3$ — in round terms a North Sea total of probably something over 1000 mrd m$^3$.

The advent of indigenous natural gas in 1967 coincided with a phase of development within the U.K. gas industry when sales were already buoyant, even though the bulk of supplies consisted of high-cost gas manufactured from

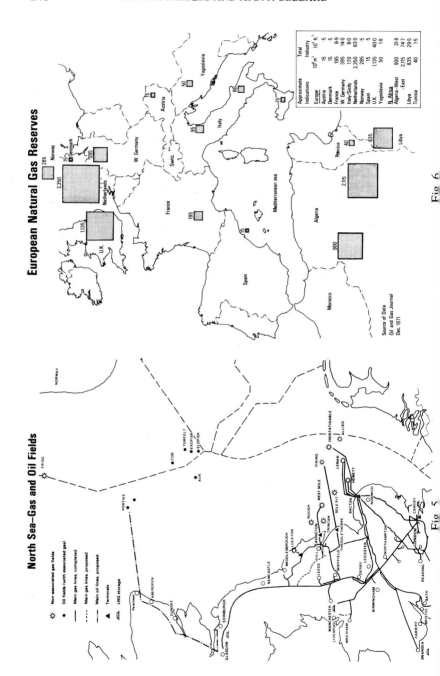

**North Sea—Gas and Oil Fields**

**European Natural Gas Reserves**

| Approximate Indications | Total $10^9 m^3$ | Industry $10^9$ n. |
|---|---|---|
| Europe | | |
| Austria | 15 | .5 |
| Denmark | 15 | .5 |
| France | 195 | 6.9 |
| W. Germany | 395 | 14.0 |
| Italy/Sicily | 170 | 6.0 |
| Netherlands | 2,350 | 83.0 |
| Norway | 285 | |
| Spain | 15 | .5 |
| U.K. | 1,135 | 40.0 |
| Yugoslavia | 50 | 1.8 |
| N. Africa | | |
| Algeria—West | 900 | 31.8 |
| —East | 2,115 | 74.7 |
| Libya | 835 | 29.5 |
| Tunisia | 40 | 1.5 |

Source of Data
Oil and Gas Journal
Dec. 1971

Fig. 6

Fig. 5

coal or oil. As might be expected, this progress has subsequently been maintained. In 1967, gas from all sources accounted for a significant part (7 per cent) of final end-users' consumption of energy, although its share in primary energy supply (reflecting the relatively small quantity of natural gas) was only 0.6 per cent. It will be seen from Fig. 4 that by 1970 natural gas had grown to provide almost 6 per cent of primary energy and the indications are that by 1975-76 this proportion will have increased to around 16 per cent if present plans are fulfilled. This represents a degree of market penetration equal to any foreseen in continental Europe, with the notable exception of the Netherlands.

Traditionally, the pattern of U.K. gas sales has concentrated on the premium market, in particular the domestic consumer. In the industrial sector gas was taken only by customers prepared to pay relatively high prices for a specialized quality fuel; bulk sales for under-boiler use were unknown. This sales pattern is now being transformed by the attention being given to the expansion of the industrial market, which is rapidly overtaking the domestic sector as the main component of gas demand.

This switch of emphasis is the logical outcome of the change in the gas industry's supply structure and reflects the crucial dual role which the bulk market must play both as a ready outlet for the considerable volumes of natural gas now available, and as the means of meeting the need for economic high load factor operation of the pipeline system. Such a rapid build-up of supplies to a new base energy market is, of course, also rewarding in terms of the present value of the revenue from accelerated sales, besides satisfying a natural requirement on the part of the government for a large indigenous energy resource to be developed with all possible despatch.

Compelling though the underlying reasons are, however, the change of marketing philosophy to one of expansion and the emergence of an important new class of gas customer in the U.K. jointly pose certain questions for the longer-term future. The demand for gas, once stimulated, is unlikely to remain static at any particular planned level, while the volume of supply from the North Sea is clearly not unlimited. The broad dimensions of the problem are illustrated below.

The original objective to market by the mid-1970s about 40 mrd m³/year of natural gas (including the LNG from Algeria) appears to be still generally valid, the implied growth being well within the capacity of the known recoverable reserves. Longer term, however, indigenous supplies appear likely to prove inadequate to meet the growing requirements of the market. On a reasonable assessment of the available data, production from known reserves plus potentials can be expected to provide a total maximum supply base of, say, 50-60 mrd m³/year by the end of the decade. Even with a relatively modest annual market growth of 5 per cent after the initial build-up period, demand should be up to this level by the early 1980s; at a higher but still not excessive rate of increase of, say, 7.5 per cent/year, a supply deficit position could

## UK Possible Natural Gas Supply and Demand Levels 1970–85

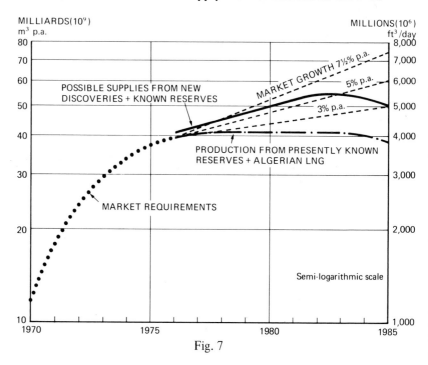

Fig. 7

emerge before 1980, leading to a substantial supply shortage by 1985. This situation is illustrated in Fig. 7.

Even if gas industry growth is throttled back in line with the requirement for energy as a whole — an increase, say, of 3-3½ per cent/year – supply difficulties will be avoided within the time span being considered only if the so-far undiscovered potential resources under the North Sea are exploited to the full. If wellhead prices remain inadequate to provide an economic return to new exploration, thereby confining the level of future supplies to the capacity of presently-known fields, a shortfall of at least 10 mrd m³ is a distinct possibility by 1985.

Enough has been said to emphasize the likelihood that, short of spectacular (and unforeseeable) new discoveries, the U.K. gas industry will, by the end of the present decade, be facing a situation where indigenous supplies of natural gas are insufficient to meet the incremental requirements of the market. Within the constraints of pipeline economics, one possible solution will be similar to that already suggested as likely in the U.S.A., namely, a more selective premium sales pattern, reverting to the traditional public distribution style, by the withdrawal of gas from certain base uses in the industrial sector where other fuels would be more appropriate. At the same time, the enhanced value of gas in the market will open the way for the importation of supplies of LNG far in excess of the existing pioneering scheme between Arzew and Canvey Island, mainly, it would appear at this stage, from either Algeria or Nigeria, as costs, availability, and other considerations permit.

## THE WORLD ENERGY PICTURE IN 1985

How will demand for total energy grow to 1985 and how will it be met? One can safely assume a close association between total energy demand and economic growth. In almost all parts of the world, economic growth is the primary objective of governments, so economic growth can be expected to continue apace, not only in the industrialized countries but also in the developing areas of the world.

For the U.S.A., which has a large, powerful, and mature economy, one can expect the maintenance of a fairly constant growth rate of around 4 per cent. Western Europe is forecast to achieve a broad growth of approximately 4.5-5.0 per cent, roughly in line with performance in the 1960s, although with some tapering-off in the mid-1970s. A slight slowing down of the Japanese economy can be expected, although the average over the period 1970-85 will still be an exceptional 9 per cent. These and other regional forecasts provide us with an estimated annual average growth in GNP for the world outside the Communist areas of around 4.5-5.0 per cent.

Whilst a crude relationship between GNP and total energy demand can be determined, it is preferable to go deeper and to identify the links between the various elements in the growth of the economy, such as industrial production and disposable incomes, and principal energy markets. However, even with the most penetrating analysis, uncertainty is inherent in any forecast, not only of economic growth itself but in its relationship to energy demand and the way in which that demand is met. Therefore, it is generally wiser to attempt to show a range of possible energy demands. Fig. 8 shows the estimated range of total energy demand, associated with the foregoing projections of economic growth, in the main regions of the world in 1985 compared with 1970. It is clear that, in general, regional patterns of energy demand will not have changed significantly from 1970. The highly industrialized nations, whose needs are increasing most in absolute terms, will continue to absorb the greater part of the world's energy resources. It is after the 1980s that the rapid growth of energy demand in the developing countries may begin to amend this picture somewhat.

The factors mentioned earlier in this paper of location and size of resources, basic cost structures, governmental intervention, environmental protection, etc., all play their role in determining how this demand will be met. Fig. 8 also shows how the mean of the range of primary energy demand may be met in 1985 compared with 1970.

It can be seen that, despite the large percentage increase in nuclear energy, the hydrocarbon fuels (oil and natural gas) will still be supplying around 70 per cent of total energy in 1985.

The main consuming regions of the world will continue to have energy deficits of various extents and to a larger degree, despite the greatest possible use of indigenous energy, consistent with cost.

Efficient use will have to be made of all fuels and the need is to utilize all forms of energy available in a way most suited to their quality, cost features, and availabilities.

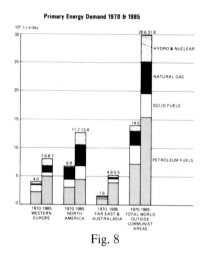

Fig. 8

## DISCUSSION

*M. J. Wells* (Amoco Europe Inc.) said that with regard to premium fuel supplies, gas did not necessarily have to be priced on a basis of equivalent Btu compared with other sources of energy, as it was a premium fuel and significant quantities could be sold irrespective of the price of other energy fuels.

For example, in Johannesburg, where coal was extremely cheap, equivalent to about 19 cents/MCF, some gas was sold there at over $1/MCF quite successfully. One could give coal away in some places and one would still sell gas. In West Germany and other West European countries some gas was sold below the fuel oil price at first.

There were now several recent examples of the realization of shortage of gas in developed countries – Western Europe and the U.S.A. The FPC in the U.S.A. had recently re-appraised its thoughts from one of adequate indigenous supplies in the U.S.A. to one of almost a panic condition.

Now significant changes in the philosophy of gas pricing, arising from a change in the future supply picture in the area of Western Europe, were apparent. In the Netherlands there was now a ban on new export contracts for Groningen gas because they realized that production potential of the field was lower than was initially estimated and foreseeable reserves had been allocated. Again, in the U.K., gas reserves were not being found so rapidly as was expected at one time.

The Algerian/U.S. East Coast LNG sales now going through approval involved a price of around 80 cents/1000 cu ft in the first year, or approximately double the cost of domestically produced gas. Any gas found in the region of 200 miles offshore in Europe in deep waters would have to obtain a similar substantial rise in price because of the increase of cost of investment in such fields.

On the question of long-distance gas supplies, one might see before long a development in trade-out arrangements between countries or parts of a country. With the existing pipeline systems in the U.S.A. and Europe and with substantial increases in the price, LNG supplies would be able to reach almost any market in those areas. But it was emphasized that the price would have to be about $1–$1.25 to get an LNG project under way.

For example, Norwegian gas could be sent to the east of the U.S.A. to replace gas supplied there from southern U.S.A. This latter source could in turn be re-directed to even the north-west U.S.A. area through trade-out shifts in supply. Again, similar shifts of gas supplies to Western Europe could be done with Russian and North African gas by agreements between those concerned without actually transporting the gas to the specific sales point.

A final point of observation was that some of the big oil companies might, more correctly, be renamed gas and oil companies the way they were developing. Some now produced almost as much gas as oil on an equivalent energy basis.

*Dr T. F. Gaskell* (British Petroleum Co. Ltd) said that if the price of North Sea gas were put up to the level of LNG costs, it was almost certain that more gas would be found in the North Sea.

*Mr Peebles*, in reply, agreed that price gave incentive but there was a limit to the amount of fossil fuel in the North Sea and recent discoveries had been oil with associated gas rather than gas alone. There was a reasonable expectation that the gas industry target of 4000 million cu ft/day would be achieved if not exceeded. However, if the gas industry wished to retain a 15 per cent share of the energy market, there would still be a deficit in our time.

*G. Laading* (Royal Norwegian Council for Scientific and Industrial Research) asked whether there was not a great deal more oil from shale and tar-sands and was there not a great deal of oil left behind in conventional reservoirs that could be produced if the price increased.

*Mr Gellard* replied that it was probable that most of the oil reserves would eventually be recovered and would result in oil being available, but there were very considerable difficulties to be overcome.

*Mr Peebles* said that work on producing the tar-sands to make natural gas was going on but it would take until 1980 and would be at a cost greater than 1000 cents/MMBtu.

*Miss M. P. Doyle* (Esso Petroleum Co. Ltd) said she worked for the world's largest energy company and had confidence in long-range forecasts. Energy included coal and nuclear, as well as all types of petroleum. It was thought that in 1972 shale would be needed. Now it was realized that shale would not come in until some time after the Alaskan oil development. It was possible that recent experience had been used too much in making forecasts and extrapolations without taking into consideration other factors that might affect the future growth rate.

*Mr Peebles* replied that they had taken in their various charts the precaution of covering various levels of demand. However, even at the lowest levels there was still a supply deficiency and confidence was just a facade. It was a valid point that in considering energy we should not put all our money on one horse. However, the paper was concerned with natural gas. The sources that they considered were ones which had been proved, whereas gas and oil from shale and gas from coal and even nuclear had not been proved on a large enough scale to be considered today. We must make forecasts on what we know and leave ourselves room to manoeuvre by considering other possibilities in the future.

# Author Index

# Subject Index